Lecture Notes in Mathematics

Edited by A. Dold and B. Eckmann

1313

Fritz Colonius

Optimal Periodic Control

Springer-Verlag

Berlin Heidelberg New York London Paris Tokyo

Author

Fritz Colonius
Institut für Dynamische Systeme, Universität Bremen Fb 3
Postfach 330440, 2800 Bremen 33, Federal Republic of Germany

Mathematics Subject Classification (1980): 49-02, 49B10, 49B27, 93-02, 34K35

ISBN 3-540-19249-2 Springer-Verlag Berlin Heidelberg New York
ISBN 0-387-19249-2 Springer-Verlag New York Berlin Heidelberg

Printing and binding: Druckhaus Beltz, Hemsbach/Bergstr.
2146/3140-543210

To Didi Hinrichsen, who taught me
how to do mathematics

CONTENTS

CHAPTER I

INTRODUCTION

1. These notes are concerned with optimal periodic control for ordinary and functional differential equations of retarded type. In its simplest version this problem can be stated as follows:

Consider a controlled system

$$\dot{x}(t) = f(x(t),u(t)), \quad t \in R_+ := [0,\infty) \tag{1}$$

where $x(t) \in R^n$, $u(t) \in R^m$. Look for a τ-periodic control function u and a corresponding τ-periodic trajectory x such that the "average cost"

$$1/\tau \int_0^\tau g(x(t),u(t))dt \tag{2}$$

is minimized (or the "average output" is maximized).
If one adds the boundary condition

$$x(0) = x(\tau), \tag{3}$$

it is sufficient to consider x and u on the compact interval $T := [0,\tau]$ only: By (3), the periodic extensions of x and u to R_+ lead to an absolutely continuous solution x of (1) on R_+.

Hence the optimal periodic control problem formulated above (abbreviated as (OPC)) is intermediate between dynamic optimization problems on R_+ and the following static or steady state optimization problem (OSS) associated with (OPC):

Minimize the "instantaneous cost"

$$g(x,u)$$

where $x \in R^n$ is a steady state corresponding to a constant control $u \in R^m$, i.e. satisfies

$$0 = f(x,u).$$

In these notes we study the relation between (OPC) and (OSS). The more complicated relation between dynamic optimization problems on R_+ and periodic problems is not considered here (some results in this direction are contained in Leizarowitz [1985], Colonius/Sieveking [1987], Colonius/Kliemann [1986]).

The fundamental problem concerning the relation between (OPC) and (OSS) can be formulated as follows: Suppose that $(x^0,u^0) \in R^n \times R^m$ is an optimal solution of (the relatively simple, finite dimensional) problem (OSS). Can the average performance be improved (in every neighborhood of the constant functions $\overline{x}^0 \equiv x^0$, $\overline{u}^0 \equiv u^0$) by allowing for τ-periodic (x,u)? That is, exist τ-periodic (x,u) satisfying the constraints of (OPC) and

$$1/\tau \int_0^\tau g(x(t),u(t))dt < g(x^0,u^0) \ ?$$

If this is the case, we call (x^0,u^0) *locally proper*.

Local properness can be tested by checking if $(\overline{x}^0,\overline{u}^0)$ satisfies necessary optimality conditions for a local solution of (OPC). If $(\overline{x}^0,\overline{u}^0)$ violates these conditions, (x^0,u^0) is locally proper. This problem becomes mathematically interesting, since first order necessary optimality conditions (for weak local) minima do *not* allow to discern steady states which are merely optimal among steady states from those which are also optimal among periodic solutions. We will prove various necessary optimality conditions for weak and strong local minima and local relaxed minima of (OPC) and develop corresponding tests for local properness. Furthermore we will relate local properness to dynamic properties of the system equation.

Other important aspects of optimal periodic control theory will not be discussed here. For existence results of optimal periodic solutions (of ordinary differential equations) we refer to Nistri [1983] and Gaines/Peterson [1983]; cp. also Miller/Michel [1980]. Numerical methods for the computation of optimal periodic solutions as well as sufficient optimality conditions are briefly reviewed in Section IX.2, below.

Actually, we consider more general system equations than (1), namely functional differential equation of retarded type

$$\dot{x}(t) = f(x_t,u(t)) \tag{4}$$

where $f: C(-r,0;R^n) \times R^m \to R^n$, $r > 0$, and $x_t \in C(-r,0;R^n)$ is given by

$$x_t(s) := x(t+s), \quad s \in [-r,0];$$

this includes delay equations of the form

$$\dot{x}(t) = f(x(t),x(t-r),u(t)),$$

where $f: R^n \times R^n \times R^m \to R^n$.

For these equations the boundary condition (3) is not adequate, since the "state" of (4) (or (5)) at time t is given by the function segment $x_t \in C(-r,0;R^n)$.

Hence (3) has to be replaced by the (infinite dimensional) condition

$$x_0 = x_\tau. \tag{6}$$

Thus we have to use the full force of optimization theory in infinite dimensional spaces in order to treat the corresponding optimal periodic problem.

2. Optimal periodic control theory was first motivated by problems from chemical engineering. Sometimes "cycling" of a chemical reactor allows to increase the average output compared to steady state operation. Here (steady state) relaxed control played as particular role, see e.g. Horn/Bailey [1968].

Early work in the field is reviewed in Bailey [1973], see also the survey Matsubara/Nishimura/Watanabe/Onogi [1981]. Other recent work includes Watanabe/Onogi/Matsubara [1981], Watanabe/Kurimoto/Matsubara [1984], Schädlich/Hoffmann/Hofmann [1983].

Besides control of chemical reactors, flight performance optimization provides a second main source of motivation. Speyer [1973,1976] observed that sometimes steady state cruise is not fuel optimal. This led to the consideration of "chattering cruise" which is a (steady state) relaxed solution (apparently, a more complete problem description avoids chattering here: Houlihan/Cliff/Kelley [1982]); see also Gilbert/Lyons [1981], Speyer/Dannemiller/Walker [1985], Chuang/Speyer [1985], Sachs/Christodopulou [1986].

Diverse other reported applications of optimal periodic control include harvesting problems (Vincent/Lee/Goh [1977], Deklerk/Gatto [1981]; cp. also Brauer [1984], Brauer/Soudack [1984]), soaring of gliders (e.g. Dickmanns [1982]), vehicle cruise (Gilbert [1976]), maintenance problems (Khandelwal/Sharma/Ray [1979]) and dynamic pricing problems (Timonen/Hämäläinen [1979]).

Early contributions to the mathematical theory of optimal periodic control were given in Horn/Lin [1967], Markus [1973], Halanay [1974] and CIME Lecture Notes edited by Marzollo [1972] (including also annotations on the history and prehistory of optimal periodic control); see also the surveys Guarbadassi [1976], Guarbadassi/Locatelli/Rinaldi [1974], Noldus [1975].

Problems with discrete system equations are considered e.g. in Bittanti/

Fronza/Guarbadassi [1974,1976], Ortlieb [1980] and Valkó/Almasy [1982].

Periodic problems with delay equations are treated in Sincic/Bailey [1978] (motivated from chemical engineering) and in Li [1985], Li/Chow [1987] (for linear equations see also Barbu/Precupanu [1978] and DaPrato [1987]).

The best available survey on optimal control of functional differential equations is still Manitius [1976] (cp. also Banks/Manitius [1974], and Oguztöreli [1966], Warga [1972], Gabasov/Kirillova [1976,1981] and for contributions from the engineering side, Koivo/Koivo [1978], Marshall [1980], Malek-Zavarei/Jamshidi [1987]). Delay equations frequently occur in chemical engineering models and ecological problems; cp. e.g. Manitius [1974] for a discussion of various models.

3. These notes are structured as follows:
Chapter II collects results from general optimization theory needed in the sequel, in particular first and second order necessary optimality conditions for problems in Banach spaces. Although Hale's book [1977] is used as a reference text for functional differential equations, Chapter III includes a sketch of duality for linear time-varying functional differential equations based on a calculus of structural operators. This allows to avoid excessive use of the Unsymmetric Fubini Theorem (which is hidden now behind the properties of the structural operators). Furthermore, extendability to the product space is discussed. Chapter IV presents a global maximum principle for strong local minima including a "stopping condition" for determination of the optimal period length. The proof relies on Ekeland's Variational Principle. Chapters V and VI contain first and second order necessary optimality conditions for weak local minima and local relaxed minima, respectively. A remarkable observation here is that - under reasonable assumptions - every ordinary optimal solution is also optimal among relaxed solutions.

The last three chapters are devoted to a discussion of local properness. Chapter VII develops tests for local properness (in particular, a so-called π-Test) which are based on the catalogue of necessary optimality conditions from Chapters IV-VI. The relation between necessary optimality conditions for the periodic and the steady state problems is discussed in detail.

In Chapter VIII, we relate local properness to dynamic properties of the system equation. We exhibit a scenario for local properness which is related to Hopf bifurcation. An example involving a retarded Liênard equation is worked out.

The final Chapter IX treats problems with ordinary differential equations. In particular, a Π-Test for problems with state constraints is proved. For a simple model of a "Continuous Flow Stirred Tank Reactor" (in Rutherford Aris' words "So sesquipedelian a style supplicates a sobriquet": we use CSTR) it is shown how local properness occurs near a Hopf bifurcation point.

In summary, our main results for optimal periodic control of functional differential equations are:

- Proof of a global maximum principle based on Ekeland's Variational Principle;
- A "stopping condition" for determination of the optimal period length;
- First and second order necessary optimality conditions for ordinary and relaxed problems with state and control constraints and isoperimetric constraints;
- Tests for local properness, in particular a Π-Test, based on the necessary optimality conditions;
- A scenario for local properness related to Hopf bifurcation;
- Discussion of two examples involving retarded Liênard equations and an ordinary differential model of a chemical reactor;

and finally

- a Π-Test for state constrained problems with ordinary differential equations.

We hope, that these results will help to renew interest in optimal periodic control theory. It is apparent from the literature cited above that a first interest in mathematical questions in this field had occurred at the beginning of the seventies. However, (1) there is continuing interest from chemical engineering and aerospace engineering; other applications, e.g. in ecology, are promising, too; (2) a further analysis of the relation between optimal periodic control and dynamic properties appears possible; (3) the results above show that periodic control of functional differential equations is much more well-behaved than control with fixed boundary values (in order to make this point clearer we have included in Chapters V and VI a discussion of fixed boundary value problems), and (4) some of the results derived here for retarded functional differential equations remain true for other infinite dimensional, in particular parabolic differential equations (cp. Colonius [1987]).

4. This research report is a revised version of my Habilitationschrift, Universität Bremen, Bremen 1986. The main revisions are (i) a sharpened and more general version of the second order necessary conditions in Section II.2 made possible by using some ideas from Werner [1984], (ii) a new proof of a stopping condition for determination of the optimal period length based on Ekeland's Variational Principle, now in Section IV.2; in this chapter, extendability to the product space is now assumed; (iii) a corrected version of a π-Test under state constraints in section IX.3.

The research reported here was performed during visits to Mathematisches Institut der Universität Graz (1983/84) and, as Visiting Assistant Professor, at Lefschetz Center for Dynamical Systems, Brown University, Providence, R.I. (1984/85). These visits were supported by a grant from Deutsche Forschungsgemeinschaft. It is a pleasure to thank Prof. F. Kappel, Universität Graz, and Prof. H.T. Banks, Brown University, for their invitations. Furthermore an invitation by the late Prof. G.S.S. Ludford, Cornell University, to take part in the Special Year on Reacting Flows was very helpful for an understanding of the CSTR problem. Prof. Matsubara, Nagoya University, draw my attention to the interesting problem of a π-Test under state constraints.

I am indebted to A.W. Manitius and D. Salamon for the permission to use some unpublished material in Chapter III. Furthermore, D. Hinrichsen and M. Brokate pointed out errors in the earlier version. Finally, I thank V. Landau, who typed the earlier version, and E. Sieber for their competent work.

5. Some remarks on the notation are in order: Standing hypotheses in a section or chapter are only repeated in statements of theorems. The end of a proof is marked by □.

For a set Q in a vector space we define the conical hull of Q with respect to $q^0 \in Q$ as

$$Q(q^0) := \{\alpha(q-q^0) : \alpha \geq 0,\ q \in Q\}.$$

The norm in a Banach space X is denoted by $|\cdot|_X$; where no confusion appears possible, we omit the index X. Furthermore, for $\rho > 0$, we let

$$X_\rho := \{x \in X : |x| \leq \rho\}.$$

The space of linear functionals on X is denoted by X', while X^* denotes the dual Banach space of bounded linear functionals on X. The dual of the space $C(a,b;R^n)$ of continuous function on [a,b] with values in R^n is identified with the space $NBV(a,b;R^n)$ of normalized functions v of bounded variation, i.e. v is left continuous on (a,b) and $v(b) = 0$. Derivatives are denoted in various ways, as it appears most convenient in the respective context. Furthermore $R_+ := [0,\infty)$.

CHAPTER II

OPTIMIZATION THEORY

This chapter collects results from general optimization theory. For most of them proofs are available in books and hence omitted here. However, complete proofs for the second order necessary conditions in section 2 are included, since the specific results we need were not available in sufficient generality. Furthermore, second order conditions play a central role in optimal periodic control theory; hence completeness in the arguments appears adequate.

After the exposition of first and second order necessary optimality conditions in sections 1 and 2, section 3 indicates a result by A.V. Fiacco on smooth dependence of optimal solutions on a parameter and cites I. Ekeland's Variational Principle.

The main results of this chapter are Theorem 1.11, Corollary 2.12 and Corollary 3.7.

1. First Order Optimality Conditions

In this section we consider the following optimization problem in Banach spaces.

Problem 1.1 Minimize $G(x)$

s.t. $F(x) \in K, \quad x \in C.$

where $G: X \to R$, $F: X \to Y$, X,Y are Banach spaces, the set $C \subset X$ is closed and convex, and $K \subset Y$ is a closed and convex cone with vertex at the origin.

For a set Q in a Banach space X define the conical hull $Q(q^o)$ of Q with respect to $q^o \in Q$ by

$$Q(q^o) := \{\alpha(q-q^o) : \alpha \geq 0,\ q \in Q\}.$$

Observe that for a convex cone K with vertex at the origin and $y^o \in K$

$$K(y^o) = \{k-\alpha y^o : \alpha \geq 0,\ k \in K\}.$$

Frequently, we abbreviate

$$Q_\rho := Q \cap X_\rho, \qquad \rho > 0.$$

The following two theorems, a generalized open mapping theorem and first order necessary optimality conditions, go back to work by S.M. Robinson [1976] (cp. also Zowe/Kurcyusz [1979], Alt [1979]). A nice, self-contained treatment is given in the lecture notes by Werner [1984].

Theorem 1.2 Let X and Y be Banach spaces and $T: X \to Y$ be a bounded linear map. Suppose that Q is a closed and convex set in X and K is a closed and convex cone with vertex at the origin in Y. Then for $q^0 \in Q$ and $y^0 \in K$ the following two statements are equivalent:

(i) $Y = TQ(q^0) - K(y^0)$

(ii) $Y_\rho \subset T(Q-q^0)_1 - K(y^0)_1$ for some $\rho > 0$.

Proof: See Werner [1984, Theorem 5.2.3].

One obtains immediately the following corollary, Werner [1984, Corollary 5.2.4].

Corollary 1.3 Suppose the hypotheses of Theorem 1.2 are satisfied. Let

$$\rho_0 := \sup\{\rho > 0: Y_\rho \subset T(Q-q^0)_1 - K(y^0)_1\}.$$

Then for $L > 1/\rho_0$ and $y \in Y$ there exist

$$\begin{Bmatrix} x \\ z \end{Bmatrix} \in L|y| \begin{Bmatrix} (Q-q^0)_1 \\ K(y^0)_1 \end{Bmatrix} \quad \text{with} \quad y = Tx + z.$$

Theorem 1.4 Let x^0 be a local minimum of Problem 1.1 and assume that the functional G is Fréchet differentiable at x^0 and the map F is continuously Fréchet differentiable at x^0. If the constraint qualification

$$F'(x^0)C(x^0) - K(F(x^0)) = Y \tag{1.1}$$

holds, then there exists $y^* \in Y^*$ satisfying

(i) $y^*y \geq 0$ for all $y \in K$

(ii) $y^*F(x^0) = 0$

(iii) $[\lambda_0 G'(x^0) - y^*F'(x^0)]x \geq 0$ for all $x \in C(x^0)$.

Proof: See Werner [1984, Theorem 5.3.2].

Remark 1.5 It suffices, naturally, that F and G are defined in a neighborhood $\mathcal{O}$ of x^0. This, being true for all following necessary optimality conditions, is very convenient if F and G are implicitly defined only.

The following corollary slightly extends the result above.

Define, for $x \in X$, $\lambda = (\lambda_0, y^*) \in R \times Y^*$, the Lagrangean functional

$$L(x,\lambda) := \lambda_0 G(x) - y^* F(x). \tag{1.2}$$

Corollary 1.6 Let the assumptions of Theorem 1.4 be satisfied and assume that either $F'(x^0)C(x^0) - K(F(x^0))$ is not dense in Y or contains a subspace of finite codimension in Y. Then there exists $0 \neq \lambda = (\lambda_0, y^*) \in R \times Y^*$ satisfying

(i) $\lambda_0 \geq 0$, $y^*y \geq 0$ for all $y \in K$

(ii) $y^*F(x^0) = 0$

(iii) $\mathcal{D}_1 L(x^0,\lambda)x \geq 0$ for all $x \in C(x^0)$.

If the constraint qualification (1.1) is satisfied, then $\lambda_0 \neq 0$.
If (1.1) is supplemented by

$$c\ell[RF(x^0) + F'(x^0)N_X + N_Y] = Y, \tag{1.3}$$

where

$$N_Y = [-K(F(x^0))] \cap K(F(x^0))$$

and

$$N_X = [-C(x^0)] \cap C(x^0)$$

are the greatest linear subspaces contained in $K(F(x^0))$ and $C(x^0)$, respectively, then, for given λ_0, the conditions (i) - (iii) above determine y^* uniquely.

Proof: If (1.1) holds, the assertion follows by Theorem 1.4. If $F'(x^0)C(x^0) - K(F(x^0))$ is not dense in Y, the assertion follows by the Hahn-Banach Theorem (e.g. Klee [1969, 1.3]). Thus it remains to discuss the case where $F'(x^0)C(x^0) - K(F(x^0))$ contains a subspace N of finite codimension in Y. By a version of the Hahn-Banach Theorem (cp. e.g. Kirsch/Warth/Werner [1978, Satz 1.1.14]) there is $y' \in Y'$ with $y'y \geq 0$ for all $y \in K$ and

$$y'F'(x^0)x \geq 0 \quad \text{for all} \quad x \in C(x^0).$$

Let M be the linear span of $B := F'(x^0)C(x^0) - K(F(x^0))$. The subspaces M and N are closed in Y and the factor space M/N is finite dimensional. We denote by $\pi: M \to M/N$ the canonical (linear and bounded) projection. Thus

$$\pi y_1 = \pi y_2 \quad \text{iff} \quad y_1 - y_2 \in N.$$

If M is a proper subspace of Y, there exists $y^* \in Y^*$ satisfying the assertions with $\lambda_0 = 0$. Thus we may assume $M = Y$. Observe that πB is a convex subset of a finite dimensional space. Thus if 0 is a boundary point of B, there exists a bounded linear functional $\bar{y}$ on M/N with

$$\bar{y}\pi F'(x^o)x \le 0 \quad \text{for all} \quad x \in C(x^o)$$

$$\bar{y}\pi y \ge 0 \quad \text{for all} \quad y \in K$$

$$\bar{y}\pi F(x^o) = 0.$$

Hence the functional $\bar{y}\pi \in Y^*$ satisfies the assertions with $\lambda_0 = 0$. Now suppose that $0 \in \text{int } B$. Then

$$Y = M = F'(x^o)C(x^o) - K(F(x^o))$$

and hence (1.1) holds.
Finally, let (1.3) be satisfied and suppose $y_1^*, y_2^* \in Y^*$ satisfy (i) - (iii) with $\lambda_0 = 0$. Then

$$(y_1^* - y_2^*)[\alpha F(x^o) + F'(x^o)x + y] = 0 \quad \text{for all} \quad \alpha \in R,\ x \in N_X,\ y \in N_Y$$

and by (1.3) $y_1^* = y_2^*$. □

Remark 1.7 Zowe/Kurcyusz [1979], Kurcyusz [1973,1976], Penot [1982], and Brokate [1980] contain more information on condition (1.1), see also Theorem 1.18, below. Condition (1.3) is very restrictive if other than equality constraints are present. Hence, in this case, one has - in general - to live with non unique Lagrange multipliers (see also Lempio/Zowe [1982]).

In the following problem, the cone constraint has a special structure which can be exploited.

Problem 1.8 Minimize $G(x)$

$$\text{s.t.} \qquad F(x) = 0, \quad H(x) \in K, \quad x \in C,$$

where $G: X \to R$, $F: X \to Y$, $H: X \to Z$, X,Y and Z are Banach spaces, C is a closed and convex subset of X, and K is a closed and convex cone in Z with vertex at the origin and *non-empty interior*.

Note that Problem 1.8 is a special case of Problem 1.1 (with cone $\{0\} \times K \subset Y \times Z$). Frequently we will refer to the constraints $F(x) = 0$ and $H(x) \in K$ as the equality and the inequality constraint, respectively, while the constraint $x \in C$ is called the explicit constraint.

Remark 1.9 In the optimal control problems considered later, the equality constraint corresponds to the system equation with boundary conditions while state and control constraints are incorporated in the inequality constraint and the explicit constraint, respectively.

Proposition 1.10 Let x^0 satisfy the constraints of Problem 1.8, and suppose that F and H are Frêchet differentiable at x^0.

Then the regularity condition (1.1) is equivalent to the following two conditions:

$$Y = F'(x^0)C(x^0) \tag{1.4}$$

$$Z = H'(x^0)[\text{Ker } F'(x^0) \cap C(x^0)] - K(H(x^0)). \tag{1.5}$$

Condition (1.5) holds in particular if there is

$$\tilde{x} \in C(x^0) \cap \text{Ker } F'(x^0) \quad \text{with} \quad H'(x^0)\tilde{x} \in \text{int } K(H(x^0)). \tag{1.6}$$

In presence of (1.5) condition (1.3) holds if

$$Y = F'(x^0)N_X \text{ and } c\ell\{RH(x^0)+H'(x^0)[\ker F'(x^0)\cap N_X] + N_Z\} = Z \tag{1.7}$$

where $N_X := [-C(x^0)] \cap C(x^0)$, $N_Z := [-K(H(x^0))] \cap K(H(x^0))$.

Proof: For Problem 1.8, condition (1.1) specializes to

$$(F'(x^0),H'(x^0))C(x^0) - \{0\}\times K(H(x^0)) = Y\times Z.$$

This, obviously, implies (1.4) and (1.5).

Conversely, let $(y,z) \in Y\times Z$ be given. By (1.4), there exists $x^1 \in C(x^0)$ with

$$y = F'(x^0)x^1.$$

By (1.5), there exist $x^2 \in C(x^0)$ and $k \in K(H(x^0))$ with

$$0 = F'(x^0)x^2$$

and

$$z - H'(x^0)x^1 = H'(x^0)x^2 - k.$$

Since $C(x^0)$ is a convex cone

$$x := x^1+x^2 \in C(x^0)$$

and one finds

$$\begin{aligned}
&(F'(x^0),H'(x^0))x - (0,k) \\
&= (F'(x^0)x^1,H'(x^0)(x^1+x^2)) - (0,k) \\
&= (F'(x^0)x^1,H'(x^0)x^1+z - H'(x^0)x^1) \\
&= (y,z).
\end{aligned}$$

Thus (1.1) follows.
Next suppose that for $\tilde{x} \in \operatorname{Ker} F'(x^0) \cap C(x^0)$ there exists a neighborhood N of $0 \in Z$ with

$$H'(x^0)\tilde{x} - N \subset K(H(x^0))$$

$$\text{i.e.} \qquad N \subset H'(x^0)\tilde{x} - K(H(x^0)).$$

Thus the cone

$$H'(x^0)[C(x^0) \cap \operatorname{Ker} F'(x^0)] - K(H(x^0))$$

contains a neighborhood of $0 \in Z$, proving (1.5).
Finally, by arguments as in the first part of this proof, one sees the last assertion.

□

We summarize first order necessary optimality conditions for Problem 1.8 in the following theorem.

Theorem 1.11 Let x^0 be a local minimum of Problem 1.8, suppose that G is Frêchet differentiable at x^0, and that F and H are continuously Frêchet differentiable at x^0. Then there exists $0 \neq \lambda = (\lambda_0, y', z^*) \in R \times Y' \times Z^*$ satisfying

(i) $\lambda_0 \geq 0$, $z^*z \geq 0$ for all $z \in K$

(ii) $z^*H(x^0) = 0$

(iii) $[\lambda_0 G'(x^0) - y'F'(x^0) - z^*H'(x^0)]x \geq 0$ for all $x \in C(x^0)$.

If $F'(x^0)C(x^0)$ is not dense in Y or contains a subspace of finite codimension in Y, then one may take $y' = y^* \in Y^*$.

If (1.4) holds, then $(\lambda_0, z^*) \neq (0,0)$. If (1.4) and (1.5) hold, then $\lambda_0 \neq 0$ and $y' = y^* \in Y^*$. If (1.6) and (1.7) hold, then for given $\lambda_0 \geq 0$, conditions (i) - (iii) above determine (y^*, z^*) uniquely.

Proof: If conditions (1.4) and (1.5) (or (1.5) and (1.6)) are satisfied, the assertions are an immediate consequence of Proposition 1.10 and Corollary 1.6. If (1.4) or (1.5) is violated, the Hahn-Banach Theorem guarantees the existence of $0 \neq \lambda = (0, y', z') \in R \times Y' \times Z'$ satisfying conditions (i) - (iii). In particular, $z'z \geq 0$ for all $z \in K$. But the cone K has non-empty interior. Thus $K - K = Z$, and z' is bounded on a neighborhood of the origin. This proves $z' \in Z^*$.

If $F'(x^0)C(x^0)$ is not dense in Y or contains a subspace of finite codimension in Y, the assertion follows as Corollary 1.6.

Finally, suppose (1.4) holds and $(\lambda_0, z^*) = (0,0)$. Then (iii) implies $y^* = 0$ contradicting non triviality. □

Remark 1.12 For Problem 1.8, the Lagrangean has the form

$$L(x,\lambda) = \lambda_0 G(x) - y^*F(x) - z^*H(x). \tag{1.8}$$

for $\lambda = (\lambda_0, y^*, z^*) \in R \times Y^* \times Z^*$. Thus condition (iii) can be written as

$$\mathcal{D}_1 L(x^0,\lambda)x \geq 0 \quad \text{for all} \quad x \in C(x^0).$$

The set of all Lagrange multipliers for Problem 1.8 is defined as

$$\Lambda(x^0) = \{0 \neq \lambda = (\lambda_0, y^*, z^*) \in R \times Y^* \times Z^*: \text{(i) - (iii) in Theorem 1.11 hold}\} \tag{1.9}$$

In the rest of this section we discuss two cases of *nested* optimization problems. The first one is important for an analysis of "local properness" in optimal *periodic* control problems (Section VII.2). The second case arises in optimal control of functional differential equations with *fixed* boundary values (Section VI.2).

Consider the following optimization problem "sitting inside" Problem 1.1.

Problem 1.13 Minimize $G(x)$

$$\text{s.t.} \quad F(x) \in K, \quad x \in \tilde{C},$$

where $\tilde{C} \subset C$.

We note the following simple result.

Proposition 1.14 Suppose that F and G have Gateaux derivatives at $x^0 \in \tilde{C}$ and assume that there exists a linear map $P: X \to X$ with

$$PC \subset \tilde{C}, \quad Px^0 = x^0$$

and that there are $\lambda_0 \geq 0$ and $y' \in Y'$ such that for all $x \in X$

$$\lambda_0 G'(x^0,x) - y'F'(x^0,x) = \lambda_0 G'(x^0,Px) - y'F'(x^0,Px).$$

Then the condition

$$\lambda_0 G'(x^0,x) - y'F'(x^0,x) \geq 0 \tag{1.10}$$

holds for all $x \in C(x^0)$ iff it holds for all $x \in \tilde{C}(x^0)$.

Proof: Suppose that the inequality (1.10) holds on $\tilde{C}(x^0)$ and let $x := \alpha(x^1 - x^0) \in C(x^0)$. Then

$$\begin{aligned} &\lambda_0 G'(x^0,x) - y'F'(x^0,x) \\ &= \lambda_0 G'(x^0,Px) - y'F'(x^0,Px) \\ &= \lambda_0 G'(x^0,\alpha(Px^1 - x^0)) - y'F'(x^0,\alpha(Px^1 - x^0)) \\ &\geq 0 \end{aligned}$$

□

<u>Remark 1.15</u> Under the assumptions of Corollary 1.6 the existence of (λ_0,y^*) satisfying (1.10) for all $x \in C(x^0)$ (resp. $x \in \tilde{C}(x^0)$) is necessary for an optimal solution x^0 of Problem 1.1 (resp. Problem 1.13). Proposition 1.14 shows that - in the considered situation - already optimality in the restricted Problem 1.13 implies the first order necessary conditions for optimality in Problem 1.1. Hence these conditions do not allow to discern between optimal solutions $x^0 \in \tilde{C}$ of Problem 1.1 and points x^0 which are merely optimal for Problem 1.13. The assumptions of Proposition 1.14 may be interpreted in the following way: Problem 1.13 can be obtained by *"projection"* of Problem 1.1.

<u>Remark 1.16</u> Let $F \equiv 0$ in Problems 1.1 and 1.13, and suppose that X is a Hilbert space, $\tilde{X}$ is a closed linear subspace of X, the set $\tilde{C}$ is given by $\tilde{C} := C \cap \tilde{X}$, and G is Frêchet differentiable at $x^0 \in \tilde{C}$. Thus $G'(x^0)$ is a continuous linear functional on X. By the Riesz Representation Theorem, $G'(x^0)$ can be identified with an element in $X = X^*$. Take P as the orthogonal projection of X onto $\tilde{X}$. Then $PC \subset \tilde{C}$ and $G'(x^0,x) = G'(x^0,Px)$ if

$$G'(x^0) \in \tilde{X} \quad \text{and} \quad PC \subset C.$$

Next suppose that $K = \{0\}$ in Problem 1.1. Hence the regularity condition (1.1) has the form

$$F'(x^0)C(x^0) = Y. \tag{1.11}$$

We show that even if (1.11) is violated (because "the cone $C(x^0)$ is too small"), we can still assure the existence of a bounded Lagrange multiplier.

Consider the following situation:

There exist Banach spaces $\tilde{X} \subset X$ and $\tilde{Y} \subset Y$ which are dense subspaces such that $C \subset \tilde{X}$ and $F(\tilde{X}) \subset \tilde{Y}$. (1.12)

The restriction of G to $\tilde{X}$ is Frêchet differentiable and F considered as a map from $\tilde{X}$ to $\tilde{Y}$ is continuously Frêchet differentiable; the set $\tilde{C}$ is convex and closed in $\tilde{X}$. (1.13)

$$F'(x^0)C(x^0) = \tilde{Y}. \tag{1.14}$$

Conditions (1.12) - (1.14) require that the assumptions of Corollary 1.6 are satisfied for the following problem sitting inside Problem 1.1.

<u>Problem 1.17</u> Minimize $G(x)$

over all $x \in \tilde{X}$ with $F(x) = 0$ and $x \in C \subset \tilde{X}$.

Thus for a local optimal solution x^0 of Problem 1.17 there exist by Corollary 1.6 Lagrange multipliers $\tilde{\lambda} = (1,\tilde{y}^*) \in R\times\tilde{Y}^*$ with

$$[G'(x^0) - \tilde{y}^*F'(x^0)]x \geq 0 \quad \text{for all} \quad x \in C(x^0). \tag{1.15}$$

The situation may be illustrated by Figure 1.

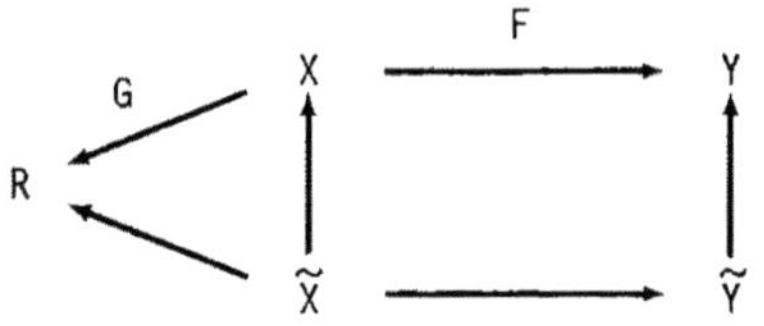

Fig. 1

Every element of Y^* can be considered as a continuous linear functional on $\tilde{Y}$, but the converse is not true. Thus the question arises, under what conditions $\tilde{y}^*$ can be continuously extended to a functional on Y.

<u>Theorem 1.18</u> Suppose that conditions (1.12) - (1.14) are satisfied for x^0 and assume additionally:

(a) $G'(x^0)$ and $F'(x^0)$ can be extended to continuous linear maps on X with values in R and Y, respectively.

(b) $F'(x^0)c\ell_X C(x^0) = Y$.

Then $\tilde{y}^*$ can be continuously extended to Y.

<u>Proof:</u> We have to show that $\tilde{y}^*$ is bounded on a ball around the origin in Y.

In order to apply the generalized Open Mapping Theorem 1.2 define

$$Q: = c\ell_X C(x^0), \quad q^0: = 0, \quad T = F'(x^0) \tag{1.16}$$

Thus Q is a closed and convex cone in X and

$$\begin{aligned} Q(q^0) &= \{\alpha(q-q^0): \alpha \geq 0,\ q \in c\ell_X C(x^0)\} \\ &= c\ell_X C(x^0) \\ &= Q - q^0. \end{aligned}$$

The assumptions of Theorem 1.2 are satisfied and assumption (b) means that (i) in Theorem 1.2 holds.

Hence there exists $\rho > 0$ such that

$$Y_\rho \subset F'(x^0)(c\ell_X C(x^0))_1.$$

Observe that for every $x \in (c\ell_X C(x^0))_1$ there is a sequence $(x_n) \subset (C(x^0))_2$ converging to x.

Hence $\tilde{y}^*$ is bounded on Y_ρ if it is bounded on $F'(x^0)(C(x^0))_2$. For $y \in F'(x^0)(C(x^0))_2$, there is $x \in C(x^0)$ with $|x|_X \leq 2$ and $y = F'(x^0)x$.
By (1.15) it follows that

$$\begin{aligned}\tilde{y}^*y &= \tilde{y}^*F'(x^0)x\\ &\leq G'(x^0)x\\ &\leq 2\|G'(x^0)\|\end{aligned}$$

where $\|G'(x^0)\|$ is the norm of $G'(x^0)$ considered as a linear functional on X. The same arguments applied to $-y$ show that y^* is bounded on Y_ρ.

□

Remark 1.19 In Theorem 1.18 it is not necessary to assume that F and G are defined on all of X.

The following example will be used in Section VI.3.

Example 1.20 Let $Q \subset R^m$ be compact and convex and define

$$C := \{u \in L^2(0,1;R^m): u(t) \in Q \text{ a.e.}\}. \tag{1.17}$$

Then C is closed and convex in $X = L^2(0,1;R^m)$.
However, for $u^0 \in C$, the cone

$$C(u^0) := \{\alpha(u-u^0): \alpha \geq 0,\ u \in C\} \tag{1.18}$$

is a proper subset of

$$\{v \in L^2(0,1;R^m): v(t) \in R_+(Q-u^0(t)) \text{a.e.}\} \tag{1.19}$$

which in turn is a subset of $c\ell C(x^0)$.

Let $\Lambda_0(x^0)$ denote the set of all *normal* Lagrange multipliers $\lambda = (1,y^*) \in R\times Y^*$; that is y^* satisfies

$$[G'(x^0) - y^*F'(x^0)]x \geq 0 \quad \text{for all} \quad x \in C(x^0). \tag{1.20}$$

Proposition 1.21 Let the assumptions of Theorem 1.18 be satisfied. Then $\Lambda_0(x^0)$ is a weakly* compact set in $R\times Y^*$.

Proof: By continuity, the inequality holds for all $x \in c\ell_X C(x^0)$. Furthermore, Theorem 1.2 implies that for some $\rho > 0$ and for all $y \in Y$ with $|y|_Y \leq \rho$ there is a $x \in c\ell_X C(x^0)$ with $|x|_X \leq 1$ and $y = F'(x^0)x$. Hence for any $y^* \in \Lambda_0(x^0)$,

$$\begin{aligned}y^*y &= y^*F'(x^0)x\\ &\leq G'(x^0)x\\ &\leq |G'(x^0)|.\end{aligned}$$

The same argument applied to $-y$ shows that $|y^*|$ is uniformly bounded. Since $\Lambda_0(x^0)$ is also weakly* closed, weak* compactness follows.

□

Remark 1.22 Proposition 1.21 shows in particular, that for $K = \{0\}$ - under the assumptions of Theorem 1.4 - the regularity condition (1.1) implies boundedness of the set of normal Lagrange mulipliers. Similar arguments can be employed to show the same assertion for general K, cp. also Zowe/Kursyusz [1979,Theorem 4.1]. Such results are important for stability considerations, see e.g. Lempio/Maurer [1980].

2. Second Order Optimality Conditions

This section proves second order necessary optimality conditions by extending an approach due to Hoffmann/Kornstaedt [1978] and Linnemann [1982]. Although only second order conditions are treated here, an analogous theory can be developed for general higher order conditions (compare also the survey Lempio/Zowe [1982])

Consider the following optimization problem.

Problem 2.1 Minimize $G(x)$

s.t. $x \in A$,

where $G: X \to R$ and A is a subset of the Banach space X.

Definition 2.2 For a subset A of a normed linear space X and $x_0 \in X$, an element $h_2 \in X$ is called a variation of order two with respect to an element $h_1 \in X$ if there exist sequences $(t_i) \subset R$, $t_i > 0$, and $(r_i) \subset X$, tending to zero, such that

$$x_i := x_0 + t_i h_1 + t_i^2 h_2 + t_i^2 r_i \in A.$$

The set of all such variations is denoted by $S^2(A,x_0,h_1)$.

Definition 2.3 Let $F: X \to Y$ be a twice Fréchet differentiable map between Banach spaces X and Y. The second variational derivative of F at x_0, $h_1 \in X$ in direction $h_2 \in X$ is

$$F^{[2]}(x_0)(h_1,h_2) = (F\circ x)''(0)/2, \tag{2.1}$$

where $x(t) := x_0 + th_1 + t^2h_2$.

Note that twice Fréchet differentiability of F at x_0 implies

$$F(x_o+th_1+t^2h_2)$$
$$= F(x_o) + tF'(x_o)h_1 + t^2F^{[2]}(x_o)(h_1,h_2) + t^2r(x_o,h_1,h_2,t) \qquad (2.2)$$

with $|r(x_o,h_1,h_2,t)| \to 0$ for $t \to 0$.

This could be taken as a starting point for defining $F^{[2]}$ without requiring Frêchet differentiability.

Proposition 2.4 Under the conditions of Definition 2.3,

$$F^{[2]}(x_o)(h_1,h_2) = F'(x_o)h_2 + 1/2F''(x_o)(h_1,h_1).$$

Proof: Follows by the chain rule (see e.g. Berger [1977,(2.1.15)]. □

We have the following second order chain rule.

Proposition 2.5 Let $F: X \to Y$, and $H: Y \to Z$ be twice continuously Frêchet differentiable between the Banach spaces X,Y and Z. Then, for $x_o,x_1,x_2 \in X$

$$(H\bullet F)''(x_o)(x_1,x_2)$$
$$= H'(F(x_o))F''(x_o)(x_1,x_2) + H''(F(x_o))(F'(x_o)x_1,F'(x_o)x_1). \qquad (2.3)$$

Proof: Follows by the ordinary chain rule. □

It is interesting to note that (2.3) is equivalent to

$$(H\bullet F)^{[2]}(x_o)(x_1,x_2) = H^{[2]}(F(x_o))(F^{[1]}(x_o)x_1,F^{[2]}(x_o)(x_1,x_2))$$

which generalizes to m-th order variational derivatives, Hoffmann/Kornstaedt [1978,Lemma 3.2].

The basic result on second order necessary optimality conditions is the following consequence of Definition 2.2.

Theorem 2.6 Let x^0 be a local minimum of Problem 2.1, and suppose that G is twice Frêchet differentiable at x^0.
Choose $h_1,h_2 \in X$ with $h_2 \in S^2(A,x^0,h_1)$ and $G'(x^0)h_1 \le 0$. Then

$$G^{[2]}(x^0)(h_1,h_2) \ge 0. \qquad (2.4)$$

Proof: By choice of h_2, there are sequences $(t_i) \subset R$, $t_i > 0$, and $(r_i) \subset X$, tending to zero, with

$$x_i := x^0 + t_i h_1 + t_i^2 h_2 + t_i^2 r_i \in A.$$

Thus twice Frêchet differentiability of G and the chain rule imply that there exist $(s_i) \subset Y$, tending to zero, with

$$G(x_i) - G(x^0) = t_i G'(x^0)h_1 + t_i^2 G^{[2]}(x^0)(h_1,h_2) + t_i^2 s_i.$$

If (2.4) is violated, it follows that

$$G(x_i) < G(x^0)$$

for i large enough, contradicting local optimality of x^0. □

In Problem 1.1, the constraint set A has the following more specific structure:

$$A_{C,F,K} := \{x \in C: F(x) \in K\}.$$

The next theorem shows how to approximate $A_{C,F,K}$. It presents a second order version of the classical Lyusternik theorem (Lyusternik [1934]):

<u>Theorem 2.7</u> Let $C \subset X$ be closed and convex, let $F: X \to Y$ be twice continuously Frêchet differentiable at $x_o \in C$ with $F(x_o) \in K$ and assume that the constraint qualification (1.1) holds. Suppose $h_1 \in C(x_o)$ with $F'(x_o)h_1 \in K(F(x_o))$. Then

$$\{h_2 \in C(x_o): F^{[2]}(x_o)(h_1,h_2) \in K(F(x_o))\} \subset S^2(A_{C,F,K}, x_o, h_1).$$

<u>Proof:</u> We first show that for all $h_1, h_2 \in X$ there are $t_o > 0$, $c_o > 0$ and maps $r: [0,t_o] \to C(x^0)$, $z: [0,t_o] \to K(F(x^0))$ with

$$\begin{pmatrix} r(t) \\ z(t) \end{pmatrix} \in c_o \left| F(x^0+th_1+t^2h_2) - F(x^0) - tF'(x^0)h_1 - t^2F^{[2]}(x^0)(h_1,h_2) \right| \begin{pmatrix} (C - x^0)_1 \\ K(F(x^0))_1 \end{pmatrix} \quad \text{for } t \in [0,t_o] \tag{2.5}$$

$$F(x^0) + tF'(x^0)h_1 + t^2F^{[2]}(x^0)(h_1,h_2) = F(x^0+th_1+t^2h_2+r(t)) - z(t) \quad \text{for } t \in [0,t_o]. \tag{2.6}$$

Define as in Corollary 1.3

$$\rho_o := \sup\{\rho > 0: Y_\rho \subset F'(x^0)(C-x^0)_1 - K(F(x^0))_1\}$$

and let $\varepsilon \in (0,\rho_0/2)$. By the mean value theorem there exists $\delta > 0$ with

$$|F(x) - F(x') - F'(x^0)(x-x')| \le \varepsilon|x-x'|$$

for all $x,x' \in x^0 + X_{3\delta}$. We may take $\delta < \rho_0/2$. Since F is twice continuously Frêchet differentiable at x^0, there is by (2.2) a $t_0 > 0$ with

$$|F(x^0+th_1+t^2h_2) - F(x^0) - tF'(x^0)h_1 - t^2F^{[2]}(x^0)(h_1,h_2)| \le \frac{\delta}{2}\rho_0$$

for all $t \in [0,t_0]$.

We may choose $t_0 > 0$ small enough such that

$$|th_1 + t^2h_2| < \delta \quad \text{for all} \quad 0 \le t \le t_0.$$

Now choose $2 > \gamma > 1$ with $\gamma(1/2 + \varepsilon/\rho_0) \le 1$ and choose a fixed $t \in [0,t_0]$.

We construct sequences $\{r_k\} \subset C(x^0)$ and $\{z_k\} \subset K(F(x^0))$ as follows:

Set $r_0 := 0$, $z_0 := 0$.

Assume r_k and z_k have already been defined.

Then by Corollary 1.3, there exist

$$\begin{Bmatrix} u_k \\ v_k \end{Bmatrix} \in \frac{\gamma}{\rho_0}|F(x^0) + tF'(x^0)h_1 + t^2F^{[2]}(x^0)(h_1,h_2) - F(x^0+th_1+t^2h_2+r_k) + z_k| \begin{Bmatrix} (C - x^0)_1 \\ K(F(x^0))_1 \end{Bmatrix}$$

with

$$F(x^0) + tF'(x^0)h_1 + t^2F^{[2]}(x^0)(h_1,h_2) - F(x^0+th_1+t^2h_2+r_k) + z^k = F'(x^0)u_k - v_k. \tag{2.7}$$

Set

$$r_{k+1} := r_k + u_k, \qquad z_{k+1} := z_k + v_k.$$

We will show that $\{r_k\}$ and $\{z_k\}$ are Cauchy sequences converging to $r = r(t)$ and $z = z(t)$ satisfying (2.5) and (2.6), respectively.

Abbreviate

$$d(t) := |F(x^0+th_1+t^2h_2) - F(x^0) - tF'(x^0)h_1 - t^2F^{[2]}(x^0)(h_1,h_2)|$$

(one has $d(t) \le \frac{\delta}{2}\rho_0$) and let

$$q := \frac{\varepsilon\gamma}{\rho_0} \quad (\le 1 - \gamma/2 < 1/2).$$

By induction we show that

(i) $$\begin{Bmatrix} r_k \\ z_k \end{Bmatrix} \in \frac{\gamma}{\rho_0}\ d(t)\ \frac{1-q^k}{1-q} \begin{Bmatrix} (C - x^0)_1 \\ K(F(x^0))_1 \end{Bmatrix}$$

(ii) $$|F(x^0) + tF'(x^0)h_1 + t^2F^{[2]}(x^0)(h_1,h_2) - F(x^0+th_1+t^2h_2+r_k) + z_k| \le d(t)q^k$$

(iii) $$\begin{Bmatrix} u_k \\ v_k \end{Bmatrix} \in \frac{\gamma}{\rho_0}\ d(t)q^k \begin{Bmatrix} (C - x^0)_1 \\ K(F(x^0))_1 \end{Bmatrix}.$$

From (i) it follows in particular that

$$|r_k| \le \frac{\gamma}{\rho_0}\ \frac{\delta\rho_0}{2}\ \frac{1}{1-q} < \delta$$

so that

$$x^0 + th_1 + t^2h_2 + r_k \in x^0 + X_{2\delta}.$$

Using also (iii) one obtains

$$x^0 + th_1 + t^2h_2 + r_k + u_k \in x^0 + X_{3\delta}.$$

For $k = 0$, assertions (i), (ii) and (iii) are true. Suppose they also hold for k. Then

(i) $$r_{k+1} = r_k+u_k \in \frac{\gamma}{\rho_0}\ d(t)\ \left(\frac{1-q^k}{1-q} + q^k\right)(C-x^0)_1 = \frac{\gamma}{\rho_0}\ d(t)\ \frac{1-q^{k+1}}{1-q}\ (C-x^0)_1$$

using the induction hypotheses (i) and (iii). The statement for z_{k+1} follows similarly.

(ii) $$|F(x^0) + tF'(x^0)h_1 + t^2F^{[2]}(x^0)(h_1,h_2) - F(x^0+th_1+t^2h_2+r_{k+1})+z_{k+1}|$$
$$= |F(x^0+th_1+t^2h_2+r_k+u_k) - F(x^0+th_1+t^2h_2+r_k) - F'(x^0)u_k|$$
$$\le \varepsilon|u_k| \le \varepsilon\ \frac{\gamma}{\rho_0}\ d(t)q^k = d(t)q^{k+1}$$

using the induction hypothesis (iii).

(iii) $$u_{k+1} \in \frac{\gamma}{\rho_0}|F(x^0) + tF'(x^0)h_1 + t^2F^{[2]}(x^0)(h_1,h_2)$$
$$- F(x^0+th_1+t^2h_2+r_{k+1}) + z_{k+1}|(C-x^0)_1 \subset \frac{\gamma}{\rho_0}d(t)q^{k+1}(C-x^0)_1$$

using (ii).

The statement for v_{k+1} follows similarly.

Assertions (i), (ii) and (iii) are thus proved and it follows that

$$\{r_k\} \subset \frac{\gamma}{\rho_0}\ d(t)\ \frac{1}{1-q}\ (C-x^0)_1 \subset \frac{2}{\rho_0}\ d(t)\ (C-x^0)_1$$

and

$$\{z_k\} \subset \frac{\gamma}{\rho_0}\ d(t)\ \frac{1}{1-q}\ (K(F(x^0))_1 \subset \frac{2}{\rho_0}\ d(t)\ K(F(x^0))_1$$

are Cauchy sequences and hence converge to

$$r = r(t) \in \frac{2}{\rho_0}\ d(t)(C-x^0)_1$$

and

$$z = z(t) \in \frac{2}{\rho_0}\ d(t)K(F(x^0))_1 \subset K(F(x^0)),\quad \text{respectively.}$$

By (iii), (u_k) and (v_k) are sequences that converge to 0, and (2.7) implies

$$F(x^0) + tF'(x^0)h_1 + t^2F^{[2]}(x^0)(h_1,h_2) = F(x^0+th_1+t^2h_2+r(t)) - z(t).$$

which proves (2.5) and (2.6) with $c_0: = 2/\rho_0$.

Now choose h_1 and h_2 as specified in the theorem, and construct $r(\cdot)$ and $z(\cdot)$ as above. Then by (2.5)

$$\frac{|r(t)|}{t^2} \le c_0\ \frac{1}{t^2}\ |F(x^0+th_1+t^2h_2) - F(x^0) - tF'(x^0)h_1 - t^2F^{[2]}(x^0)(h_1,h_2)|$$

and hence by (2.2) $\frac{r(t)}{t^2} \to 0$ for $t \to 0$.

It remains to show that $x^0+th_1+t^2h_2+r(t) \in A_{C,F,K}$. We have $h_i \in C(x^0)$, i.e. $h_i = \alpha_i(c_i-x_0)$ with $\alpha_i \ge 0$, $c_i \in C$ for $i = 1,2$. By (2.5)

$$r(t) = \alpha(t)(x(t)-x^0)$$

with $0 \le \alpha(t)$, $\lim_{t\to 0^+} \alpha(t) = 0$ and $c(t) \in C$. Since C is convex

$$x^0 + th_1 + t^2h_2 + r(t)$$
$$= (1-\alpha_1 t-\alpha_2 t^2-\alpha(t))x^0 + \alpha_1 tc_1 + \alpha_2 tc_2 + \alpha(t)c(t) \in C(x^0)$$

provided that $1 - \alpha_1 t - \alpha_2 t^2 - \alpha(t) > 0$. This is the case for all sufficiently small $t > 0$.

Similarly from $F'(x^0)h_1 \in K(F(x^0))$ it follows that

$$F'(x^0)h_1 = \alpha(k-F(x^0)) \quad \text{with} \quad \alpha \ge 0,\quad k \in K.$$

By (2.5)

$$z(t) = \alpha(t)(k(t)-F(x^0))$$

with $\alpha(t) \ge 0$, $\lim_{t\to 0^+} \alpha(t) = 0$, $k(t) \in K$.

Thus (2.6) implies

$$F(x^0 + th_1 + t^2h_2 + r(t))$$
$$= F(x^0) + F'(x^0)h_1 + t^2F^{[2]}(x^0)(h_1,h_2) + z(t)$$
$$= (1 - \alpha t - \alpha(t))F(x^0) + \alpha tk + t^2F^{[2]}(x^0)(h_1,h_2) + \alpha(t)k(t)$$
$$\in K(F(x^0))$$

for t small enough, since $K(F(x^0))$ is a convex cone. □

<u>Remark 2.8</u> For general higher derivatives (but $K = \{0\}$, $C = X$) this result is due to Hoffmann/Kornstaedt [1982,Theorem 4.3] (cp. also Ben-Tal/Zowe [1982,Proposition 7.2]). The proof above follows closely the proof for first order approximation given in Werner [1984,Theorem 5.2.5].

Using this approximation result one obtains the following second necessary optimality condition.

<u>Theorem 2.9</u> Let x^0 be a local minimum of Problem 1.1 and assume that the functional G is twice Fréchet differentiable at x^0 and the map F is twice continuously Fréchet differentiable at x^0. Assume that the constraint qualification (1.1) is satisfied. Choose $h_1,h_2 \in X$ satisfying

$$G'(x^0)h_1 \le 0 \tag{2.8}$$
$$F'(x^0)h_1 \in K(F(x^0)),\ F^{[2]}(x^0)(h_1,h_2) \in K(F(x^0)) \tag{2.9}$$
$$h_1 \in C(x^0),\ h_2 \in C(x^0). \tag{2.10}$$

Then $G^{[2]}(x^0)(h_1,h_2) \ge 0$.

<u>Proof:</u> immediate from Theorem 2.6 and Theorem 2.7. □

The variations h_1,h_2 allowed here are very special. In the following theorem, more general variations are allowed, since Lagrange multipliers are introduced in the optimality conditions.

<u>Theorem 2.10</u> Let x^0 be a local minimum of Problem 1.1, and assume that the functional G is twice Fréchet differentiable at x^0 and the map F is twice continuously Fréchet differentiable at x^0. Assume that either $F'(x^0)C(x^0) - K(F(x^0))$ is not dense in Y or contains a subspace of finite codimension in Y. Choose $h \in X$ with

$$G'(x^0)h \le 0,\quad F'(x^0)h \in K(F(x^0)),\quad h \in C(x^0) \tag{2.11}$$

Then there exists $0 \neq \lambda = (\lambda_0,y^*) \in R_+ \times Y^*$ having the following properties:

$$y^*y \geq 0 \quad \text{for all} \quad y \in K, \quad y^*F(x^0) = 0 \tag{2.12}$$

$$y^*F'(x^0)h = 0 \tag{2.13}$$

$$\lambda_0 G'(x^0)h' - y^*F'(x^0)h' \geq 0 \quad \text{for all} \quad h' \in C(x^0). \tag{2.14}$$

$$[\lambda_0 G''(x^0) - y^*F''(x^0)](h,h) \geq 0. \tag{2.15}$$

Proof: If $F'(x^0)C(x^0) - K(F(x^0))$ is not dense in Y, there exists $0 \neq y^* \in Y^*$ with

$$y^*[F'(x^0)C(x^0) - K(F(x^0))] = 0.$$

Thus the assertion is valid with $\lambda_0 = 0$ and either y^* or $-y^*$ depending on the sign of

$$[\lambda_0 G''(x^0) - y^*F''(x^0)](h,h) = -y^*F''(x^0)(h,h).$$

If $F'(x^0)C(x^0) - K(F(x^0))$ contains a subspace of finite codimension but (1.1) is violated, the assertion follows similarly (cp. the proof of Corollary 1.6).

Hence we may assume that (1.1) holds. Let $h_1 := h$ and define the following convex sets in $R \times Y$:

$$W_1 := R_- \times K(F(x^0))$$

$$W_2 := \{(G,F)^{[2]}(x^0)(h_1,h_2)) : h_2 \in C(x^0)\} - W_1.$$

First we show that W_2 has non void interior. By (1.1) and the Open Mapping Theorem 1.2, there exists a neighborhood V_X of $0 \in X$ s.t.

$$F'(x^0)(C(x^0) \cap V_X) - K(F(x^0))$$

is a neighborhood of $0 \in Y$. Choose $\tilde{r} < 0$. Then there is a neighborhood V_R of $0 \in R$ with

$$\tilde{r} + V_R - G'(x^0)V_X \subset R_-$$

where, if necessary, V_X is replaced by a smaller neighborhood of $0 \in X$, again denoted by V_X.

We have constructed the following neighborhood U of $0 \in R \times Y$

$$U = V_R \times [F'(x^0)(C(x^0) \cap V_X) - K(F(x^0))].$$

Now let $\tilde{h}_2 := 0$. We show that

$$(G,F)^{[2]}(x^0)(h_1,\tilde{h}_2) - (\tilde{r},0) \in \text{int } W_2.$$

Let $(-r, F'(x^0)h_2 - y)$ be an arbitrary element of U. Then

$$(G,F)^{[2]}(x^0)(h_1,\tilde{h}_2) - (\tilde{r},0) + (-r, F'(x^0)h_2 - y)$$

$$= (G'(x^o)(\tilde{h}_2+h_2) + \tfrac{1}{2}\, G''(x^o)(h_1,h_1),\ F'(x^o)(\tilde{h}_2+h_2) + \tfrac{1}{2}\, F''(x^o)(h_1,h_1))$$
$$- (\tilde{r} + r - G'(x^o)h_2, -y)$$
$$= (G,F)^{[2]}(x^o)(h_1,\tilde{h}_2+h_2) - (\tilde{r} + r - G'(x^o), -y) \in W_2.$$

Thus $\text{int}\, W_2 \neq \emptyset$.

Furthermore, Theorem 2.9 implies

$$\text{int}\, W_2 \cap W_1 = \emptyset,$$

since an element in this intersection would lead to h_1, h_2 satisfying (2.8) - (2.10) and

$$G^{[2]}(x^o)(h_1,h_2) < 0.$$

Hence, by Eidelheit's separation theorem (see e.g. Luenberger [1968, Theorem 3,p.133] or Werner [1984,Theorem 3.2.4]) there exist $0 \neq \lambda = (\lambda_o, -y^*) \in R \times Y^*$ and $r \in R$ with

$$\lambda w \leq r \quad \text{for all} \quad w \in W_1 \tag{2.16}$$

$$\lambda w \geq r \quad \text{for all} \quad w \in W_2. \tag{2.17}$$

These relations are satisfied with $r = 0$ since W_1 is a cone with vertex at the origin. Now (2.13) follows immediately from (2.16), and (2.17) implies

$$\lambda_o G^{[2]}(x^o)(h_1,h_2) - y^* F^{[2]}(x^o)(h_1,h_2) \geq 0 \tag{2.18}$$

for all $h_2 \in C(x^o)$. Thus also (2.14) and (2.15) follow. Finally, (2.13) is a consequence of (2.14), (2.12) and (2.11). □

Note that in the theorem above, the Lagrange muliplier λ may depend on the variation h. The following proposition indicates a condition which ensures that λ may be chosen independent of h.

<u>Proposition 2.11</u> Suppose that in addition to the assumptions of Theorem 2.10, that (1.1) holds and that for every $h \in X$ satisfying (2.11) there exists $x \in C(x^o)$ with $G'(x^o)x - y^* F'(x^o)x = 0$ and

$$F^{[2]}(x^o)(h,x) \in K(F(x^o)).$$

Then there exists $y^* \in Y^*$ such that (2.12) - (2.15) hold with $\lambda_o = 1$ for *all* $h \in C(x^o)$.

<u>Proof:</u> Fix $\hat{h}$ and $\bar{h}$ satisfying (2.11) and choose corresponding $\hat{y}^*$ and $\bar{y}^*$ such that (2.12) - (2.15) are satisfied. Let $\bar{x} \in X$, $\bar{k} \in K$,

$\bar{\alpha} \in R$ be such that

$$F^{[2]}(x^o)(\bar{h},\bar{x}) = \bar{k} + \bar{\alpha}\, F(x^o) \in K(F(x^o)). \tag{2.19}$$

We only have to show that (2.15) holds for $h = \bar{h}$ and $\lambda = (1,\mathcal{y}^*)$.
Using (2.12) - (2.14) one computes

$$\begin{aligned} & G''(x^o)(\bar{h},\bar{h}) - \mathcal{y}^*F''(x^o)(\bar{h},\bar{h}) \\ &= G''(x^o)(\bar{h},\bar{h}) + 2\mathcal{y}^*[\bar{k} + \bar{\alpha}F(x^o) - F'(x^o)\bar{x}] \\ &\geq G''(x^o)(\bar{h},\bar{h}) - 2\mathcal{y}^*F'(x^o)\bar{x} \\ &\geq G''(x^o)(\bar{h},\bar{h}) - 2G'(x^o)\bar{x} \\ &= G''(x^o)(\bar{h},\bar{h}) - 2\bar{y}^*F'(x^o)\bar{x} \\ &= G''(x^o)(h,h) - \bar{y}^*F''(x^o)(\bar{h},\bar{h}) \\ &\geq 0 \end{aligned}$$

□

Next we apply the results above to Problem 1.8 and summarize second order necessary optimality conditions for later reference.

Recall the definition of the Lagrangean L and of the set $\Lambda(x^o)$ of Lagrange multipliers for Problem 1.8 in Remark 1.12.

<u>Corollary 2.12</u> Let x^o be a local minimum of Problem 1.8, assume that the functional G is twice Frêchet differentiable at x^o and that the maps F and H are twice continuously Frêchet differentiable at x^o. Suppose further that $F'(x^o)C(x^o)$ contains a subspace of finite codimension in Y.

Then $\Lambda(x^o) \neq \emptyset$ and for every $h \in X$ with

$$G'(x^o)h \leq 0, \quad F'(x^o)h = 0, \quad H'(x^o)h \in K(H(x^o)), \quad h \in C(x^o) \tag{2.20}$$

there exist $\lambda = (\lambda_o,y^*,z^*) \in \Lambda(x^o)$ with

$$z^*H'(x^o)h = 0 \tag{2.21}$$

$$\mathcal{D}_1L(x^o,\lambda)h' \geq 0 \quad \text{for all} \quad h' \in C(x^o) \tag{2.22}$$

$$\mathcal{D}_1\mathcal{D}_1L(x^o,\lambda)(h,h) \geq 0 \tag{2.23}$$

If (1.4) and (1.6) hold then $\lambda_o \neq 0$.

<u>Proof:</u> follows by Theorem 1.11, Theorem 2.10 and Proposition 1.9.

<u>Remark 2. 13</u> If $C = X$ and $H \equiv 0$, the condition $G'(x^o)h \leq 0$ may be omitted, since every $h \in C(x^o)$ with $F'(x^o)x = 0$ satisfies, by the first order conditions, also $G'(x^o)x = 0$.

3. Further Results

In this section we formulate a result due to A.V. Fiacco on stability of optimal solutions under parameter changes and cite Ekeland's Variational Principle.

Consider the following family of optimization problems depending on a parameter $\alpha \in R^k$.

Problem 3.1^α Minimize $G(x,\alpha)$ over $x \in R^n$

s.t. $F(x,\alpha) = 0$

$H(x,\alpha) \in R^q_-$

where $G, F = (F^i)$, and $H = (H^j)$ are functions from an open subset of $R^n \times R^k$ into R, R^p, and R^q, respectively, and the last condition means that $H(x,\alpha)$ is an element of the natural negative cone in R^q.

We are interested in the behaviour of a local solution $x^o(\alpha)$ of Problem 3.1^α for α close to α^o.

The following second order sufficient conditions for a strict (but not necessarily isolated, i.e. locally unique) minimum will be needed for the desired stability result.

Theorem 3.2 Let x^o satisfy the constraints of Problem 3.1^{α_o} and suppose that $F(x,\alpha_o)$ and $H(x,\alpha_o)$ are twice continuously differentiable with respect to x at x^o. Furthermore suppose that

there exist Lagrange multipliers $y = (y^i) \in R^p$ and $z = (z^j) \in R^q$ such that (3.1)

$$z^T H(x^o),\alpha^o) = 0 \quad \text{and} \quad z \in R^m_+$$

and $L_x(x^o,\alpha^o) = 0$,

where $L(x,\alpha) := G(x,\alpha) - y^T F(x,\alpha) - z^T H(x,\alpha)$ and the following second order condition is satisfied:

$$h^T L_{xx}(x^o,\alpha_o)h > 0 \quad \text{for all} \quad 0 \neq h \in R^n \quad \text{satisfying} \qquad (3.2)$$

$$G_x(x^o,\alpha_o)h = 0$$

$$F_x(x^o,\alpha_o)h = 0$$

$$H^j_x(x^o,\alpha_o)h \leq 0 \quad \text{if} \quad H^j(x^o) = 0.$$

Then x^o is a strict local minimum of Problem 3.1^{α_o}.

Proof: See e.g. Han/Mangasarian [1979].

If we add constraint qualifications to the assumptions above, we get the desired *smooth* stability result.

Theorem 3.3 Suppose that $x^0 \in R^n$ satisfies the constraints of Problem$^{\alpha_o}$ and conditions (3.1) and (3.2). Assume additionally

The functions F, G and H are twice continuously differentiable in x and F, G, F_x, G_x, and H_x are continuously differentiable in α in a neighborhood of (x^o,α_o) (3.3)

The gradients $F^i_x(x^o,\alpha_o)$, $i = 1,\ldots,p$, and $H^j_x(x^o,\alpha_o)$ with $H^j(x^0,\alpha_o) = 0$, are linearly independent and $z^j > 0$ if $H^j(x^o,\alpha_o) = 0$. (3.4)

Then

(i) x^0 is a isolated local minimum of Problem 3.1^{α_o} and the Lagrange multipliers y and z are unique,

(ii) there is a continuously differentiable function $(x(\alpha),y(\alpha),z(\alpha)) \in R^n{\times}R^p{\times}R^q$ defined on a neighborhood of α_o such that $(x(\alpha_o),y(\alpha_o),z(\alpha_o)) = (x^0,y,z)$ and $(x(\alpha),y(\alpha),z(\alpha)$ satisfy conditions (3.1) and (3.2) with α instead of α_o.

(iii) The point $x(\alpha)$ is an isolated local minimum of Problem 3.1^{α} for α in a neighborhood of α_o.

This theorem is due to Fiacco [1976]. The development above follows closely Fiacco [1983,Section 3.2], where a lot more information on related results is given; cp. also Bank/Guddat/Klatte/Kummer/Tammer [1982].

Remark 3.4 For α close to α_o, the set of active inequality constraints (i.e. $H^j(x^0,\alpha) = 0$) is not changed, strict complementary slackness (i.e. the second condition in (3.2)) holds and the regularity condition in the first part of (3.4) remains valid.

Remark 3.5 Smooth stability for optimization problems in Banach spaces (with equality constraints only) is established in Ioffe/Tikhomirov [1979]. For results on *continuous* stability in more general situations see S.M. Robinson [1976] and W. Alt [1979,1983].

Finally, we formulate Ekeland's Variational Principle which refers to points, "almost" minimizing a given function and, roughly, states that there is a "nearby point" which actually minimizes a slightly perturbed functional.

Theorem 3.6 Let V be a complete metric space with associated metric d, and let $F: V \to \mathbb{R} \cup \{\infty\}$ be a lower semi-continuous functional which is bounded below. If u is a point in V satisfying

$$F(u) \leq \inf F + \varepsilon$$

for some $\varepsilon > 0$, then there exists a point v in V such that

(i) $F(v) \leq F(u)$

(ii) $d(u,v) \leq \varepsilon$

(iii) For all $w \neq v$ one has $F(w) + \varepsilon d(w,v) > F(v)$.

This result is due to Ekeland [1974]. See also Clarke [1983], Aubin/Ekeland [1984]. Besides diverse other applications, Ekeland [1979], this result has become an important tool in optimal control theory mainly due to Clarke's work.

Corollary 3.7 Let V,F be as in the previous theorem. Let $\varepsilon > 0$, $u \in V$ such that

$$F(u) \leq \inf F + \varepsilon.$$

Then there exists $v \in V$ such that

(i) $F(v) \leq F(u)$

(ii) $d(u,v) \leq \sqrt{\varepsilon}$

(iii) For all $w \neq v$ in V one has $F(w) > F(v) - \sqrt{\varepsilon}\, d(w,v)$.

Proof: Use Theorem 3.6 with the distance

$$d_\varepsilon(u,v) := d(u,v)/\sqrt{\varepsilon}$$

(see also Ekeland [1979,p.456]). □

CHAPTER III

RETARDED FUNCTIONAL DIFFERENTIAL EQUATIONS

As a general reference for retarded functional differential equations we use Hale's book [1977]. However, this chapter gives some complements and cites certain details.

The first two sections follow very closely Colonius/Manitius/Salamon [1985] and develop the structure and duality theory for linear retarded equations in the state space of continuous functions and in the product space. The proofs are only sketched. The results of section 2 are only needed in Chapter IV.

The third section contains local and global existence and uniqueness results for nonlinear equations. Furthermore, a theorem on Hopf bifurcation which is needed for illustrative purposes in Chapter VIII, is cited.

The main results of this chapter are Proposition 1.4, Theorem 1.6, Proposition 3.2 and Theorem 3.3.

1. Structure Theory of Linear Equations

We consider the linear time varying retarded functional differential equation

$$\dot{x}(t) = L(t)x_t, \quad t \in R \tag{1.1}$$

where $x(t) \in R^n$ and x_t is defined by

$$x_t(s) := x(t+s), \quad -r \le s \le 0, \quad 0 < r < \infty,$$

and $L(t): C(-r,0;R^n) \to R^n$ and we assume

There exists a function $m \in L^1_{\ell oc}(-\infty,\infty)$ such that (1.2)
for a.a. $t \in R$ and all $\varphi \in C(-r,0;R^n)$
$|L(t)\varphi| \le m(t)|\varphi|_\infty$
and for every $\varphi \in C(-r,0;R^n)$, the function $t \to L(t)\varphi$, $t \in R$, is measurable.

Lemma 1.1 There exists a $n\times n$ matrix valued function $\eta(t,\tau)$ with the following properties:

(i) For every $t \in R$ the function $\eta(t,\cdot) \in NBV(-r,0;R^n)$, i.e. $\eta(t,\cdot)$ is of bounded variation and normalized in the sense that $\eta(t,\tau)$ is left continuous in τ for $-r < \tau < 0$, $\eta(t,\tau) = 0$ for $\tau \geq 0$ and $\eta(t,\tau) = \eta(t,-r)$ for $\tau \leq -r$.

(ii) For a.a. $t \in R$ and all $\varphi \in C(-r,0;R^n)$
$L(t)\varphi = \int_{[-r,o]} [d_\tau \eta(t,\tau]\varphi(\tau)$

(iii) The function $\eta(\cdot,\cdot)$ is measurable on $R\times R$.

Proof: Existence of η with the properties (i) and (ii) follows by the Riesz representation theorem. Measurability of $\eta(\cdot,\cdot)$ can be shown using Bourbaki [1968,Chapter V,Exercice 6]. □

A function $x(\cdot) \in C(t_1-r,t_2,R^n)$ is said to be a solution of (1.1) if it is absolutely continuous and satisfies (1.1) a.e. By Hale [1977, Chapter 6], equation (1.1) admits a unique solution on $[t_1,t_2]$ for every initial condition of the form

$$x(t_1+\tau) = \varphi(\tau), \quad -r \leq \tau \leq 0, \tag{1.3}$$

where $\varphi \in C(-r,0;R^n)$ and for an additional forcing term in $L^1(t_1,t_2;R^n)$.

Correspondingly, one can consider the solution segment $x_t \in C(-r,0;R^n)$ as the state of (1.1) at time t. The evolution of this state determines a family $\Phi(t,s)$, $t \geq s$, of bounded linear operators on $C(-r,0;R^n)$ defined by

$$\Phi(t,t_1)\varphi = x_t \in C(-r,0;R^n)$$

where $x(t)$, $t \geq t_1-r$, is the unique solution of (1.1) and (1.3). Then $\Phi(t,s)$ is a well-defined strongly continuous evolution operator whose properties are specified in the following proposition (see Hale [1977]).

Proposition 1.2 Let (1.2) be satisfied. Then

(i) $\Phi(t,s)$ is a bounded linear operator on $C(-r,0;R^n)$ for all $t,s \in R$, $t \geq s$.

(ii) $\Phi(t,t) = Id$ for all $t \in R$

(iii) $\Phi(t,s)\Phi(s,\tau) = \Phi(t,\tau)$ for $t \ge s \ge \tau$.

(iv) For every compact interval $[t_1,t_2] \subset R$ there exists a constant M such that $\Phi(t,s) \le M$ for $t_1 \le s \le t \le t_2$.

(v) $\Phi(t,s)$ is strongly continuous on the domain $\{(t,s) \in R^2 | t \ge s\}$ i.e. $\Phi(t,s)\varphi$ is a continuous function for every $\varphi \in C(-r,0;R^n)$.

(vi) $\Phi(t,s)$ is compact for $t \ge s+r$; thus $Id - \Phi(t,s)$ is a Fredholm operator for $t \ge s+r$.

Integration of (1.1) yields

$$x(t_1+s) = f^{t_1}(s) + \int_0^s \int_{[-\sigma,0]} [d_\tau \eta(t_1+\sigma,\tau)] x(t_1+\sigma+\tau) d\sigma, \quad s \ge 0, \tag{1.4}$$

where $f^{t_1}(\cdot) \in C(0,t_2-t_1;R^n)$ is given by

$$f^{t_1}(s) = \varphi(0) + \int_0^s \int_{[-r,-\sigma]} [d_\tau \eta(t_1+\sigma,\tau)] \varphi(\sigma+\tau) d\sigma, \quad s \ge 0. \tag{1.5}$$

Note that $f^{t_1}(s)$ is constant for $s \ge r$.

The forcing term f^{t_1} may be used to define an alternative state concept for the retarded equation (1.1). Consider (1.1) at $t = t+s$ and split the integral in terms depending on values of $x(\cdot)$ with an argument less, resp. greater than t. One obtains

$$x(t+s) = f^{t_1}(t+s-t_1) + \int_0^{t+s-t_1} \int_{[-\sigma,t-\sigma-t_1)} [d_\tau \eta(t_1+\sigma,\tau)] x(t_1+\sigma+\tau) d\sigma \tag{1.6}$$

$$+ \int_0^{t+s-t_1} \int_{[t-\sigma-t_1,0]} [d_\tau \eta(t_1+\sigma,\tau)] x(t_1+\sigma+\tau) d\sigma$$

$$= f^t(s) + \int_0^s \int_{[-\sigma,0]} [d_\tau \eta(t+\sigma,\tau) x(t+\sigma+\tau) d\sigma, \quad s \ge 0,$$

where $f^t(\cdot) \in C(0,r;R^n)$ is given by

$$f^t(s) = f^{t_1}(t+s-t_1) + \int_0^{t+s-t_1} \int_{[-\sigma,t-\sigma-t_1)} [d_\tau \eta(t_1+\sigma,\tau)] x(t_1+\sigma+\tau) d\sigma \tag{1.7}$$

for $0 \le s \le r$ and again $f^t(s) = f^t(r)$ for $s > r$.

Note that the shifted forcing term $f^t(\cdot)$ contains all the information from the past history of the solution at time t which is needed to determine the future behaviour of the solution $x(t+s)$, $s \ge 0$. This forcing function $f^t(\cdot) \in C(0,r;R^n)$ can be considered as the state of equation (1.4) at time $t \ge t_1$. The evolution of this state determines the family $\Psi(t,s)$, $t \ge s$, of bounded linear operators on $C(0,r;R^n)$ defined by

$$\Psi(t,t_1)f^{t_1} = f^t \in C(0,r;R^n) \tag{1.8}$$

where $x(t)$, $t \ge t_1$ is the unique solution of (1.4) and $f^t(\cdot)$ is defined by (1.7).

The properties of $\Psi(t,s)$ are indicated in the following proposition.

Proposition 1.3 Let (1.2) be satisfied. Then

(i) $\Psi(t,s)$ is a bounded linear operator on $C(0,r,R^n)$ for all $t,s \in R$ with $t \ge s$.

(ii) $\Psi(t,t) = Id$ for all $t \in R$.

(iii) $\Psi(t,s)\Psi(s,\tau) = \Psi(t,\tau)$ for $t \ge s \ge \tau$.

(iv) For every compact interval $[t_1,t_2]$ there exists a constant M such that $|\Psi(t,s)| \le M$ for $t_1 \le s \le t \le t_2$.

Proof: Introduce the operator T on $X: = C(0,t_2-t_1;R^n)$ by

$$[Tx](s): = \int_0^s \int_{[-\sigma,0]} [d_\tau \eta(t_1+\sigma,\tau)]x(\sigma+\tau)d\sigma.$$

Using the norm

$$\|x(\cdot)\|_\gamma := \sup_{0 \le s \le t_2-t_1} |x(s)|e^{-\gamma s}$$

($\gamma > 0$ small enough), which is equivalent to the sup-norm, one shows that T is a contraction, hence $I-T$ is boundedly invertible. This shows that (1.4) is uniquely solvable for all $f^{t_1} \in C(0,t_2-t_1;R^n)$. These arguments prove (i) and (iv); (ii) is obvious and (iii) follows by straightforward computation. □

The relation between the two state concepts can be described by two structural operators $F(t): C(-r,0;R^n) \to C(0,r;R^n)$ and $G(t): = C(0,r;R^n) \to C(-r,0;R^n)$. The operator $F(t_1)$ maps the initial function $\varphi \in C(-r,0;R^n)$ of (1.1), (1.3) into the corresponding forcing term $f^{t_1}(\cdot) \in C(0,r;R^n)$ of (1.4) which is given by (1.5) and the operator $G(t_1)$ maps this forcing term $f^{t_1}(\cdot)$ into the corresponding solution segment $x_{t_1+r} \in C(-r,0;R^n)$ of (1.4) at time t_1+r. These two operators can be described explicitly by the formulae:

$$[F(t)\varphi](s) = \varphi(0) + \int_0^s \int_{[-r,-\sigma]} [d_\tau \eta(t+\sigma,\tau)]\varphi(\sigma+\tau)d\sigma, \tag{1.9}$$

$$[G(t)^{-1}\varphi](s) = \varphi(s-r) - \int_0^s \int_{[-\sigma,0]} [d_\tau \eta(t+\sigma,\tau)]\varphi(\sigma+\tau-r)d\sigma, \qquad (1.10)$$

for $0 \le s \le r$ and $\varphi \in C(-r,0;R^n)$.

The operator $G(t)^{-1}$ is boundedly invertible and its inverse is the desired operator $G(t)$.

Proposition 1.4 Let (1.2) be satisfied. Then

(i) The operator $G(t)\colon C(0,r;R^n) \to C(-r,0;R^n)$ is bijective.

(ii) $\Phi(t+r,t) = G(t)F(t)$, $\Psi(t+r,t) = F(t+r)G(t)$.

(iii) $F(t)\Phi(t,s) = \Psi(t,s)F(s)$, $\Phi(t+r,s+r)G(s) = G(t)\Psi(t,s)$, $t \ge s$.

Proof: Statement (i) has been shown in the preceeding proof. The first equation in (ii) is an immediate consequence of the definitions of the operators $F(t)$ and $G(t)$. The second equation in (ii) and the assertions in (iii) follow by straightforward computation. □

The equation $F(t)\Phi(t,t_1) = \Psi(t,t_1)F(t_1)$ can be interpreted in the following way. If $f^{t_1}(\cdot)$ is given by (1.5), if $x(t)$, $t \ge t_1-r$, is the unique solution of (1.1), (1.3), and if $f^t(\cdot)$, $t \ge t_1$, is defined by (1.6), then $f^t = F(t)x_t$.

Next we describe the duality theory for equation (1.1). Recall that the dual space of $C(a,b;R^n)$ is identified with $NBV(a,b;R^n)$ by means of the duality pairing

$$\langle g,\varphi\rangle = \int_a^b [d_\tau g^T(\tau)]\varphi(\tau) \qquad (1.11)$$

$g \in NBV(a,b;R^n)$, $\varphi \in C(a,b;R^n)$.

We note the following formulae for the adjoint operators

$F^*(t)\colon NBV(0,r;R^n) \to NBV(-r,0;R^n)$ and

$G^*(t)\colon NBV(-r,0;R^n) \to NBV(0,r;R^n)$.

Lemma 1.5 Let $\psi \in NBV(0,r;R^n)$ be given. Then for $-r \le \tau < 0$ the following equations hold:

$$[F^*(t)\psi](\tau) = \psi(0) - \int_0^r [\eta^T(t+s,\tau-s) - \eta^T(t+s,-s)]\psi(s)ds \qquad (1.12)$$

$$[G^*(t)^{-1}\psi](\tau) = \psi(\tau+r) + \int_\tau^0 \eta^T(t+r+\sigma,\tau-\sigma)\psi(\sigma+r)d\sigma. \qquad (1.13)$$

Proof: Follows by computation using the Unsymmetric Fubini Theorem (cp. e.g. Johnson [1984]). □

The operators $F^*(t)$ and $G^*(t)$ are related to the *transposed equation*

$$z(t) - z(t_2) = - \int_t^{t_2+r} [\eta^T(\alpha,t-\alpha)-\eta^T(\alpha,t_2-\alpha)]z(\alpha)d\alpha, \quad t \le t_2. \tag{1.14}$$

This equation is sometimes called the "formal adjoint equation" and has been used in the theory of functional differential equations for a long time (see Henry [1971] and Hale [1977]).

Equation (1.14) admits a unique solution $z(t)$ in the space $NBV(t_1,t_2+r;R^n)$, $t_1 < t_2$, for every final condition of the form

$$z(t_2+s) = \psi(s), \quad 0 \le s \le r, \tag{1.15}$$

where $\psi \in NBV(0,r;R^n)$ (see e.g. Hale [1977,p.148,Theorem 3.1]). This motivates the definition of the state of system (1.14) at time $t \le t_2$ to be the solution segment $z^t \in NBV(0,r;R^n)$ given by

$$z^t(s) = \begin{cases} z(t+s), & 0 \le s < r \\ 0 & , \quad s = r \end{cases} \tag{1.16}$$

Equation (1.14), (1.15) can be rewritten in the form

$$z(t_2+\tau) = g_{t_2}(\tau) - \int_\tau^0 \eta^T(t_2+\sigma,\tau-\sigma)z(t_2+\sigma)d\sigma, \quad \tau < 0, \tag{1.17}$$

where $g_{t_2}(\cdot) \in NBV(-r,0;R^n)$ is given by

$$\begin{aligned} g_{t_2}(\tau) &= \psi(0) - \int_0^r [\eta^T(t_2+\sigma,\tau-s) - \eta^T(t_2+s,-s)]\psi(s)ds \\ &= [F^*(t_2)\psi](s), \quad -r \le \tau < 0. \end{aligned} \tag{1.18}$$

This shows that the functional analytic adjoint of the forcing term operator is the forcing term operator for the transposed equation.

Comparing the formulae (1.17) and (1.13) one sees that a function $z(\cdot) \in NBV(t_2-r,t_2;R^n)$ satisfies (1.17) iff

$$z^{t_2-r} = G^*(t_2-r)g_{t_2}. \tag{1.19}$$

where $z^{t_2-r} \in NBV(0,r;R^n)$ is given by (1.16). Since $G^*(t_2-r)$ is bijective, this shows that equation (1.17) admits a unique solution for every $g_{t_2} \in NBV(-r,0;R^n)$. As for equation (1.1) one may now define the forcing term g_{t_2} to be the final state of equation (1.17). The corresponding state at time $t \le t_2$ can be obtained by means of a time shift as for (1.6). The shifted equation takes the form

$$z(t+\tau) = g_t(\tau) - \int_\tau^0 \eta^T(t+\sigma,\tau-\sigma)z(t+\sigma)d\sigma), \quad \tau < 0, \tag{1.20}$$

where $g_t(\cdot) \in NBV(-r,0;R^n)$ is given by

$$g_t(\tau) = g_{t_2}(t-t_2+\tau) - \int_\tau^{t_2} \eta^T(\alpha,t+\tau-\alpha)z(\alpha)d\alpha, \quad -r \le \tau < 0. \tag{1.21}$$

The forcing term g_t of the shifted equation is now regarded as the state of system (1.17) at time $t \le t_2$.

The following theorem establishes the duality of equations (1.1) and (1.14).

Theorem 1.6 (i) Let $\psi \in NBV(0,r;R^n)$ be given, let $z(t)$, $t \le t_2+r$ be the corresponding solution of (1.14), (1.15), and let $z^t \in NBV(0,r;R^n)$ be defined by (1.16). Then

$$z^t = \Psi^*(t_2,t)\psi, \quad t \le t_2. \tag{1.22}$$

(ii) Let $g(\cdot) = g_{t_2} \in NBV(-r,0;R^n)$ be given, let $z(t)$, $t \le t_2$, be the corresponding solution of (1.17), and let $g_t \in NBV(-r,0;R^n)$ be defined by (1.21). Then

$$g_t = \Phi^*(t_2,t)g_{t_2}, \quad t \le t_2. \tag{1.23}$$

Proof: Follows again using the Unsymmetric Fubini Theorem. □

Remark 1.7 The equation

$$F^*(t)\Psi^*(t_2,t) = \Phi^*(t_2,t)F^*(t_2) \tag{1.24}$$

can now be interpreted in the following way: If $g_{t_2}(\cdot) \in NBV(-r,0;R^n)$ is given by (1.18), if $z(t)$, $t \le t_2+r$, is the unique solution of (1.14), (1.15), and if $g_t(\cdot) \in NBV(-r,0;R^n)$ is given by (1.21), then $g_t = F^*(t)z^t$.

In addition to equation (1.1) and (1.14) we also consider the inhomogeneous equations

$$\dot{x}(t) = L(t)x_t + f(t), \quad \text{a.a. } t \in T := [t_1,t_2] \tag{1.25}$$

with $f \in L^1(T,R^n)$ and

$$z(t) - z(t_2) = - \int^{t_2+r} [\eta^T(\alpha,t-\alpha)-\eta^T(\alpha,t_2-\alpha)]z(\alpha)d\alpha \tag{1.26}$$
$$+ \int_{[t,t_2]} [d\mu^T(\alpha)]g(\alpha), \quad \text{a.a.} \quad t \in T$$

where $\mu = (\mu_i)$, $i = 1,\dots,n$, μ_i are regular Borel measures on T and $g \in C(T,R^n)$.

Define for $\alpha \in R^n$

$$(X_0\alpha)(s) := \begin{cases} 0 & -r \le s < 0 \\ \alpha & s = 0 \end{cases} \tag{1.27}$$

$$(Y_0\alpha)(s) := \begin{cases} -\alpha & s = 0 \\ 0 & 0 < \alpha \le r. \end{cases} \tag{1.28}$$

Observe that for all $\psi \in C(-r,0;R^n)$, $\alpha \in R^n$, $t \in T$

$$\alpha^T\psi(0) = \int_{-r}^{0} [d_s(F^*(t)Y_0\alpha)(s)]^T\psi(s) \tag{1.29}$$

(note $X_0\alpha \notin NBV(-r,0;R^n)$ for $\alpha \neq 0$, but $Y_0\alpha \in NBV(0,r;R^n)$).

We have the following variation of constants formulae.

Proposition 1.8 (i) Let $x(t)$, $t \ge t_1-r$, be the unique solution of (1.25), (1.3). Then

$$x_t = \Phi(t,t_1)\varphi + \int_{t_1}^{t} \Phi(t,s)X_0f(s)ds, \quad t \ge t_1.$$

(ii) Let $z(t)$, $t \le t_2+r$ be the unique solution of (1.26),(1.15). Then

$$z^t = \Psi^*(t,t_2)\psi + \int_{[t,t_2]} [d\mu^T(s)]\Psi^*(s,t)Y_0g(s), \quad t \le t_2.$$

Remark 1.9 In (i) above, φ may have a jump at t_1. The integrals in (i) (and (ii)) have to be interpreted in R^n; that is, each function is first evaluated in $s \in [-r,0]$ and then the integral is obtained as a usual integral in R^n (cp. Hale [1977,p.146]).

Remark 1.10 The state concept based on the forcing function has first been introduced by Miller [1974] for the description of Volterra integro-differential equations. Independently, it was introduced by Bernier/Manitius [1978], Manitius [1980], Delfour/Manitius [1980] for time invariant retarded functional differential equations in the product space framework (i.e. in $M^2 = R^n \times L_2(-r,0;R^n)$ instead of $C(-r,0;R^n)$) on the basis of a so called structural operator F. Diekmann [1981] used this concept in $C(-r,0;R^n)$. For generalizations see Salamon [1982], Delfour/Karrakchou [1987], Diekmann/Van Gils [1984]. Delfour [1977] treated time-varying systems with constant delays. The forcing function state concept has also been useful in certain numerical approximation schemes, Salamon [1985], Lasiecka/Manitius [1985].

Remark 1.11 Define for $L: C(-r,0;R^n) \to R^n$ the transposed operators $L^T: C(0,r;R^n) \to R^n$ by $L^T\varphi := \int_{-r}^{0} d\eta^T(s)\varphi(-s)$ (cp. Lemma 1.1). Then (1.14) reduces to

$$\dot{z}(t) = - L^T z^t = - \int_{-r}^{0} d\eta^T(s) z(t-s).$$

2. Extentability to the Product Space

In this section we consider the functional differential equation (1.1) in the product space $M^p(-r,0;R^n) = R^n \times L^p(-r,0;R^n)$, that is, we want to allow for initial conditions of the form

$$x(t_1) = \varphi^0, \quad x(t_1+\tau) = \varphi^1(\tau), \quad -r \le \tau \le 0, \tag{2.1}$$

where $\varphi = (\varphi^0,\varphi^1) \in M^p(-r,0;R^n)$. In addition to (1.2) the following fundamental extendability hypothesis is needed in order to give a meaning to the right hand side of equation (1.1).

Let $1 \le p < \infty$.

For all $-\infty < t_1 < t_2 < \infty$ there exists $k \in L^q(t_1-r,t_2;R)$, $1/p + 1/q = 1$, such that (2.2)

$$\int_{t_2}^{t_2} |L(t)x_t|dt \le \left(\int_{t_1-r}^{t_2} |k(t)|^q dt\right)^{1/q} \left(\int_{t_1-r}^{t_2} |x(t)|^p dt\right)^{1/p}$$

for all $x \in C(t_1-r,t_2;R^n)$.

This condition is e.g. satisfied for time-varying equations with constant delays or with distributed delays. A deeper analysis is given in Colonius/Manitius/Salamon [1985, Section 4].

Consider $\varphi \in M^p(-r,0;R^n)$ to be the initial state of (1.1) and define the state at time $t > 0$ to be the pair

$$\hat{x}(t) = (x(t),x_t) \in M^p(-r,0;R^n). \tag{2.3}$$

The time evolution of this state of (1.1) can be described by an extended evolution operator $\Phi_M(t,t_o)$ on the state space $M^p[-r,0;R^n]$ as we will see below. Correspondingly we have the natural injection ι of $C[-r,0;R^n]$ into $M^p[-r,0;R^n]$ which maps φ into $\iota\varphi = (\varphi(0),\varphi)$.

In order to extend the forcing function state concept to the product space consider the integrated equation (1.5) with the forcing term $f^{t_1}(\cdot) \in L^p_{\ell oc}(0,\infty;R^n)$ given by

$$f^{t_1}(s) = \varphi^0 + \int_0^s \int_{[-r,-\sigma]} [d_\tau \eta(t_1+\sigma,\tau)]\varphi^1(\sigma+\tau)d\sigma, \quad s > 0. \tag{2.4}$$

with $\varphi^1(\tau) := 0$ for $\tau \notin [-r,0]$. Note that the function $f^{t_1}(s)$ defined by (2.4) is absolutely continuous on $[0,r]$ and constant for $s > r$. We will consider the integrated equation (1.4) with more general forcing terms in $L^p_{\ell oc}[0,\infty,R^n]$ which are constant for $s > r$. These can be identified with pairs $f = (f^0,f^1) \in M^p[0,r;R^n]$ via

$$f^{t_1}(s) = \begin{cases} f^1(s), & 0 \le s < r, \\ f^0 \quad, & s > r. \end{cases} \tag{2.5}$$

We consider the pair $f \in M^p[0,r;R^n]$ to be the initial state of equation (1.4), (2.5) and define the state at time $t > t_1$ to be the pair

$$\hat{f}(t) = (f^t(r),f^t) \in M^p(0,r;R^n) \tag{2.6}$$

where $f^t(\cdot) \in L^p_{\ell oc}(0,\infty;R^n)$ is the forcing term of the shifted equation (1.6) given by

$$f^t(s) = \begin{cases} f^1(t+s-t_1) + \int_0^{t+s-t_1} \int_{[-\sigma,t-\sigma-t_1)} [d_\tau \eta(t_1+\sigma,\tau)]x(t_1+\sigma+\tau)d\sigma, & 0 \le s < r, \\ f^0 + \int_0^{t+r-t_1} \int_{[-\sigma,t-\sigma-t_1)} [d_\tau \eta(t_1+\sigma,\tau)]x(t_1+\sigma+\tau)d\sigma, & s > r. \end{cases}$$

Note that this expression is obtained by inserting (2.5) into (1.7). We will see below that the evolution of the forcing function state $\hat{f}(t)$ of (1.4) can be described by an extended evolution operator $\Psi_M(t,t_0)$ on $M^p(0,r;R^n)$. Furthermore, the relation between the initial function $\hat{x}(t)$ and the forcing function state $f(t)$ leads naturally to extended structural operators $F_M(t)$ and $G_M(t)$. More precisely, one has the following relations

$$\begin{aligned} \hat{x}(t) &= \Phi_M(t,s)\hat{x}(s), \\ \hat{f}(t) &= \Psi_M(t,s)\hat{f}(s), \\ \hat{f}(t) &= F_M(t)\hat{x}(t), \quad \hat{x}(t+r) = G_M(t)\hat{f}(t). \end{aligned} \tag{2.7}$$

The following results assure that all the expressions in the above equations are well defined and that there exist unique solutions of (1.1), (2.1) or, respectively, (1.4).

Lemma 2.1 Suppose that (1.2) and (2.2) are satisfied. Then the following statements hold.

(i) For every $\varphi \in M^p(-r,0;R^n)$ there exists a unique solution $x(\cdot) \in L^p(t_1-t,r_2;R^n)$ of (1.1), (2.1) which is absolutely continuous on $[t_1,t_2]$ and depends continuously on φ.

(ii) For $t_1 \le s \le t$ the operators $F(t)$ and $\Phi(t,s)$ given by (1.9) and (1.10), respectively, admit unique continuous extensions

$$F_M(t): M^p(-r,0;R^n) \to M^p(0,r;R^n)$$

and

$$\Phi_M(t,s): M^p(-r,0;R^n) \to M^p(-r,0;R^n)$$

satisfying

$$\begin{aligned} \iota F(t) &= F_M(t)\iota \\ \iota\Phi(t,s) &= \Phi_M(t,s)\iota \end{aligned} \tag{2.8}$$

(iii) The extended operators are uniformly bounded in the region $t_1 \le s \le t \le t_2-r$.

Proof: Let $\varphi \in C(-r,0;R^n)$. Define $x: [t-r,t+r] \to R^n$ by $x(s) = \varphi(s-t)$ for $s \in [t-r,t]$ and $x(s) = 0$ for $s \in (t,t+r]$.

$$\begin{aligned} |F(t)\Phi|_{W^{1,1}} &= |\varphi(0)| + \int_0^r \Big| \int_{[-r,-\sigma)} [d_\tau\eta(t+\sigma,\tau)]\varphi(\sigma+\tau)\Big| d\sigma \\ &= |\varphi(0)| + \int_0^r \Big| \int_{[-r,0]} d_\tau\eta(t+\sigma,\tau)x(t+\sigma+\tau)\Big| d\sigma \\ &= |\varphi(0)| + \int_t^{t+r} |L(s,x_s)| ds \end{aligned}$$

and, by using (2.2)

$$\begin{aligned} &\le |\varphi(0)| + \|k\|_q \|x\|_p \\ &\le c|(\varphi(0),\varphi)|_{M^p} \quad \text{for some } c > 0. \end{aligned}$$

This shows that for $t \in [t_1,t_2-r]$, $F(t)$ has a unique continuous linear extension to $M^p[-r,0;R^n] \to W^{1,1}[0,r;R^n]$. Then the existence, uniqueness and continuity of $F_M(t)$ follow by continuity of the natural embedding of $W^{1,1}$ into M^p for all $1 < p < \infty$.

Now consider equation (1.4) with forcing term f^{t_1} given by (2.4) with $\varphi \in (-r,0;R^n)$. This forcing term is in $W^{1,1}(t_1,t_2;R^n)$ and one can conclude that equation (1.4) has a unique solution $x(\cdot) \in L^p(t_1,t_2;R^n)$

depending ontinuously on $\varphi \in M^p(-r,0;R^n)$. This proves (i).

The operator $\Phi_M(t,t_1)$ defined by

$$\Phi_M(t,t_1)\Phi = (x(t),x_t)$$

where $x(\cdot)$ is the unique solution of (1.1), (2.1) satisfies the remaining assertions of the lemma. □

To extend the operators $G(t)$ and $\psi(t,s)$ to the space $M^p(0,r;R^n)$ consider equation (1.4) with arbitrary forcing terms in $L^p_{\ell oc}(0,\infty;R^n)$.

<u>Lemma 2.2</u> Suppose that the conditions (1.2) and (2.2) are satisfied. Then the following statements hold:

(i) For every $f^{t_1} \in L^p_{\ell oc}(0,\infty;R^n)$ there exists a unique solution $x(\cdot) \in L^p_{\ell oc}(t_1,\infty;R^n)$ of (1.4) depending continuously on f^{t_1}.

(ii) For $t_1 \le s \le t \le t_2-r$ the operators $G(t)$ and $\psi(t,s)$ given by (1.9) and (1.10) respectively admit unique continuous extensions $G_M(t)\colon M^p(0,r;R^n) \to M^p(-r,0;R^n)$ and $\Psi_M(t,s)\colon M^p(0,r;R^n) \to M^p(0,r;R^n)$ satisfying

$$\begin{aligned} \iota G(t) &= G_M(t)\iota \\ \iota\Psi(t,s) &= \Psi_M(t,s). \end{aligned} \tag{2.9}$$

(iii) The extended operators are uniformly bounded in the region $t_1 \le s \le t \le t_2-r$.

<u>Proof:</u> Statement (i) follows by a modification of the proof of Propositon 1.3. In order to prove (ii) observe that if $f^{t_1}(s)$ is continuous for $s > t_1+r$. This allows us to define

$$G_M(t_1)f = (x(t_1+r),x_{t_1+r})$$

$$\Psi_M(t,t_1)f = \hat{f}(t)$$

for $f \in M^p(0,r;R^n)$, where $x(\cdot) \in L^p(t_1,t_2;R^n) \cap C(t_1+r,t_2;R^n)$ is the unique solution of (1.4), (2.5) and $\hat{f}(t) = (f^t(r),f^t) \in M^p(0,r;R^n)$ is given by (2.6).

Now $x(\cdot)$ depends continuously on f and it can be shown as in Lemma 2.1 that $\hat{f}(t)$ depends continuously on $x(\cdot)$ and f. These operators satisfy the remaining assertions of the lemma. □

<u>Proposition 2.3</u> Suppose that (1.2) and (2.2) are satisfied. Then the

extended operators $F_M(t)$, $G_M(t)$, $\Phi_M(t,s)$, $\psi_M(t,s)$ satisfy properties analogous to those stated in Propositions 1.2 - 1.4.

Proof: All the statements follow from the fact that ιC is dense in M^p and all the operators satisfy uniform bounds by Lemma 2.1, 2.2.

□

We will show that under hypotheses (1.2) and (2.2) the integral adjoint equation (1.14) can be considered in the state space $W^{1,q}[0,r;R^n]$.

This naturally leads us to restrict the state space $NBV[0,r;R^n]$ of equation (1.14) to the space $W^{1,q}(0,r;R^n)$. More precisely consider the injection $\iota^*: W^{1,q}[I,r;R^n] \to NBV[0,r;R^n]$ given by

$$(\iota^*\psi)(s) = \begin{cases} \psi(s) & 0 \le s < r \\ 0 & s = r. \end{cases} \tag{2.10}$$

An analogous injection can be defined for functions defined on $[-r,0]$.

We are given the natural duality pairing between the spaces $C(0,r;R^n)$ and $NBV(0,r;R^n)$ and the injections ι and ι^*. Requiring that ι^* be a dual operator of ι in the functional analytic sense forces us to identify the dual space of $M^p(0,r;R^n)$ with $W^{1,q}(0,r;R^n)$ via the duality pairing

$$\langle\psi,f\rangle_{W^{1,q},M^p} = -\psi^T(r)f^0 + \int_0^r \dot{\psi}^T(s)f^1(s)ds, \quad p^{-1} + q^{-1} = 1. \tag{2.11}$$

Similarly, we identify the dual space of $M^p(-r,0;R^n)$ with $W^{1,q}(-r,0;R^n)$ via the duality pairing

$$\langle g,\varphi\rangle_{W^{1,q},M^p} = -g^T(0)\varphi^0 + \int_{-r}^0 \dot{g}^T(\tau)\varphi^1(\tau)d\tau.$$

These identifications have the advantage, that the results above on extendability to the product spaces can be directly translated via duality into results on restrictability of the adjoint equation to the Sobolev space $W^{1,q}$. Each of the operators $F(t)$, $G(t)$, $\Phi(t,s)$, $\Psi(t,s)$ has a continuous extension to the corresponding product spaces iff their dual operators $F^*(t)$, $G^*(t)$, $\Phi^*(t,s)$, $\Psi^*(t,s)$ restrict to bounded linear operators on the corresponding $W^{1,q}$ spaces. In particular, under the conditions (1.2) and (2.2) one has the existence of

$\Phi_M^*(t,s) \in L(W^{1,q}(-r,0;R^n))$, $\Psi^*(t,s) \in L(W^{1,q}(0,r;R^n)$,

$F_M^*(t) \in L(W^{1,q}[0,r;R^n),W^{1,q}(-r,0;R^n))$,

$G_M^*(t) \in L(W^{1,q}(-r,0;R^n),W^{1,q}(0,r;R^n))$ satisfying

$$\iota^*\Phi^*_M(t,s) = \Phi^*(t,s)\iota^*$$
$$\iota^*\Psi_M(t,s) = \Psi^*(t,s)\iota^*$$
$$\iota^*F^*_M(t) = F^*(t)\iota^* \tag{2.12}$$
$$\iota^*G^*_M(t) = G^*(t)\iota^*$$

This means that the adjoint equation (1.14) can in fact be studied in the state space $W^{1,1}(0,r;R^n)$.

Corollary 2.4 Suppose that the conditions (1.2) and (2.2) are satisfied with $1 \le p < \infty$, let $1/p + 1/q = 1$ and let $t_1 \le t_2$ be given. Then for every $\psi \in W^{1,q}(t_1,t_2+r;R^n)$ the unique solution $z(\cdot)$ of (1.14), (1.15) lies in $W^{1,q}(t_1,t_2+r;R^n)$ and depends in this space continuously on ψ.

Remark 2.5 One can in fact show that under the assumptions (1.2) and (2.2) the adjoint equation (1.14) can by written as a functional differential equation (Colonius/Manitius/Salamon [1985,Theorem 3.6]).

3. Nonlinear Equations

We consider the nonlinear retarded functional differential equation

$$\dot{x}(t) = f(x_t,t), \tag{3.1}$$

where $x_t(s) := x(t+s)$, $s \in [-r,0]$, $0 < r < \infty$.

Assume that equation (3.1) satisfies the following conditions.

> The function $f: \mathcal{O}_C \times T \to R^n$, where $\mathcal{O}_C \subset C(-r,0;R^n)$ is open and $T := [t_1,t_2]$, $-\infty < t_1 < t_2 < \infty$, is continuously Fréchet differentiable with respect to the first and measurable with respect to the second argument; there exists a function $q: R_+ \times T \to R_+$ such that $q(s,\cdot) \in L^1(T,R)$ for all $s \in R_+$, $q(\cdot,t)$ is monotonically increasing for a.a. $t \in T$ and that for all $\varphi \in \mathcal{O}_C$ and a.a. $t \in T$ (3.2)
>
> $|f(\varphi,t)| + |\mathcal{D}_1 f(\varphi,t)| \le q(|\varphi|_\infty,t)$
>
> here $|\mathcal{D}_1 f(\varphi,t)|$ denotes the operator norm of $\mathcal{D}_1 f(\varphi,t): C(-r,0;R^n) \to R^n$.

The Caratheodory and Lipschitz conditions in (3.2) imply local existence and uniqueness of solutions (Hale [1977,Section 2.6]). However, we are always interested in solutions defined on the whole interval T. The following proposition indicates a sufficient condition for the existence of global solutions.

Proposition 3.1 Suppose that in addition to (3.2) the following condition is satisfied:

There exists a constant $c > 0$ such that for all $\varphi \in C(-r,0;R^n)$ and a.a. $t \in T$ (3.3)

$$\varphi(0)^T f(\varphi,t) \le c(1+|\varphi|^2).$$

Then for every $\varphi \in C(-r,0;R)$ there exists a unique function $x(\cdot) \in C(t_1-r,t_2;R^n)$ which is absolutely continuous on $[t_1,t_2]$ and satisfies (2.1) almost everywhere.

Proof: Suppose that $x(t)$ is the local solution of (3.1) corresponding to φ. Then

$$\begin{aligned} 1/2 \frac{d}{dt} |x(t)|^2 &= x(t)^T \dot{x}(t) \\ &= x(t)^T f(x_t,t) \\ &\le c(1+|x_t|^2). \end{aligned}$$

Thus

$$\begin{aligned} |x_t|^2 &\le |\varphi|^2 + 2c \int_1^t (1+|x_s|^2)ds \\ &= |\varphi|^2 + 2c(t-t_1) + \int_{t_1}^t |x_s|^2 ds. \end{aligned}$$

Now Gronwall's inequality implies

$$\begin{aligned} |x_t|^2 \le\ & |\varphi|^2 + 2c(t-t_1) \\ & + [|\varphi|^2 + 2c(t-t_1)]\exp(t-t_1). \end{aligned}$$

By (3.2) and the mean value theorem the function f is completely continuous in φ. Thus global existence on T follows (see Hale [1977,p.43,Theorem 3.2]). □

If one is interested in necessary conditions for an optimal solution of a control problem, one starts with a global (optimal) solution x^o.

Thus we assume now that the initial value problem for (3.1) with

$$x_{t_1} = \varphi^o \qquad (3.4)$$

where $\varphi^0 \in \mathcal{O}_C$, has a global solution x^0 on $[t_1-r,t_2]$, with $x_t^0 \in \mathcal{O}_C$, for all $t \in T = [t_1,t_2]$.

Assuming (3.2), we can define

$$F: C(T;R^n)\times C(-r,0;R^n) \to C(T;R^n)$$

by

$$F(x,\varphi)(t) := \varphi(0) + \int_{t_1}^{t} f(x_s,s)ds, \tag{3.5}$$

where

$$x_{t_1}(s) := \varphi(s), \quad s \in [-r,0];$$

this is well-defined for (x,φ) in a neighborhood of $(x^0|T,\varphi^0)$.

The function F is continuously Fréchet differentiable and in a neighborhood of (x^0,φ^0), the equation

$$x = F(x,\varphi) \tag{3.6}$$

is equivalent to (3.1), (3.4).

Proposition 3.2 Suppose that equation (3.1) satisfies (3.2), and x^0 is a solution of (3.1) on $T := [t_1,t_2]$ with $x_{t_1}^0 = \varphi^0 \in C(-r,0;R^n)$. Then

(i) There exists a unique continuously Fréchet differentiable function S, defined in a neighborhood of φ^0 such that $S(\varphi)$ is the unique solution of (3.6).

(ii) The derivative $x = S'(\varphi^0)\varphi$ is the unique solution of

$$x_{t_1} = \varphi, \quad \dot{x}(t) = \mathcal{D}_1 f(x_t^0,t)x_t, \quad \text{a.a. } t \in T.$$

Proof: The proof uses the implicit function theorem and the mean value theorem (see e.g. Berger [1977,Theorems 3.1.10 and 2.1.19]) and Lebesgue's Theorem on dominated convergence. In order to show Fréchet differentiability one first computes the Gateaux derivative and observes that the derivative is linear and continuous. Hence Fréchet differentiability follows (Berger [1977,Theorem 2.1.13]). Continuous Fréchet differentiability follows by (2.2). Furthermore, $Id - \mathcal{D}_1F(x^0,u^0)$ is an isomorphism on $C(T;R^n)$ by the results in section 1. □

Finally, we cite a result on Hopf bifurcation for retarded functional differential equations of the form

$$\dot{x}(t) = f(x_t,\alpha), \quad t \geq 0 \tag{3.7}$$

where $f: C(-r,0;R^n)\times R \to R^n$.

Let there exist $x^\alpha \in R^n$, $\alpha \in R$ with $f(\bar{x}^\alpha,\alpha) = 0$, where $\bar{x}^\alpha$ is the constant function $\bar{x}^\alpha(s) \equiv x^\alpha$.

Suppose that the maps f and $\alpha \to x^\alpha$ are C^3 and that for $\alpha = \alpha_0$ the linearized equation

$$\dot{x}(t) = \mathcal{D}_1 f(\bar{x}^{\alpha_0},\alpha_0)x_t, \quad t \geq 0 \tag{3.8}$$

has a simple purely imaginary eigenvalue $z_0 = j\omega_0 \neq 0$, i.e.

$$\text{rank}\ \Delta(j\omega_0,\alpha_0) = \text{rank}[Idj\omega_0 - \mathcal{D}_1 f(\bar{x}^{\alpha_0},\alpha_0)(e^{j\omega_0 \cdot})] = n-1$$

and all eigenvalues $z_j \neq \pm z_0$ of (3.8) satisfy $z_j \neq kz_0$, $k \in Z$.

Since $\alpha \to \mathcal{D}_1 f(\bar{x}^\alpha,\alpha)$ is continuously Fréchet differentiable, Hale [1977,Lemma 2.2,p.171] implies that there is an open interval containing α_0, such that the equation

$$\dot{x}(t) = \mathcal{D}_1 f(\bar{x}^\alpha,\alpha)x_t, \quad t \geq 0 \tag{3.9}$$

has a simple eigenvalue $z(\alpha)$ with $z(\alpha_0) = j\omega_0$ and $z(\alpha)$ has a continuous derivative $z'(\alpha)$ at $\alpha = \alpha_0$. We say that a Hopf bifurcation occurs at $\alpha = \alpha_0$ if the conditions stated above are satisfied and $\text{Re}\ z'(\alpha_0) > 0$.

In fact, Hale [1977,Theorem 1.1,p.246] yields the following result (cp. also Hassard/Kazarinoff/Wang [1981]).

<u>Theorem 3.3</u> Suppose that a Hopf bifurcation occurs at $\alpha = \alpha_0$ in equation (3.7). Then there are constants $a_1 > 0$, $\alpha_1 > 0$, $\delta_1 > 0$ and functions $\alpha(a)$, $\omega(a)$, and a $\omega(a)$-periodic function $x^*(a)$ with $\alpha(0) = \alpha_0$ and $\alpha(a)$, $\omega(a)$, and $x^*(a)$ being continuously differentiable in a for $|a| < a_1$ such that $x^*(a)$ is a solution of

$$\dot{x}(t) = f(x_t,\alpha(a)), \quad t \geq 0. \tag{3.10}$$

Furthermore, for $|\alpha-\alpha_0| < \alpha_1$, $|\omega-\omega_0| < \delta_1$, every ω-periodic solution x of equation (3.10) with $|x_t-\bar{x}^\alpha| < \delta_1$ must be of this type.

CHAPTER IV

STRONG LOCAL MINIMA

The purpose of this chapter is to prove a global maximum principle for optimal periodic control of functional differential equations. The proof is based on Ekeland's Variational Principle. For ordinary differential equations, the result reduces to Pontryagin's Maximum Principle (with periodic boundary conditions). Also a condition referring to optimal choice of the period is given.

The main results of this chapter are Theorem 2.1 and Theorem 2.2.

1. Problem Formulation

We consider the following optimal periodic control problem for retarded functional differential equations.

Problem 1.1 Minimize $1/(t_2-t_1) \int_{t_1}^{t_2} g(x(t),u(t),t)dt$

$$\text{s.t.} \quad \dot{x}(t) = f(x_t,u(t),t), \quad \text{a.a. } t \in T: = [t_1,t_2]$$

$$x_{t_1} = x_{t_2}$$

$$u(t) \in \Omega \quad \text{a.a. } t \in T,$$

where $\mathcal{O}_x \subset R^n$, $\mathcal{O}_u \subset R^m$ and $\mathcal{O}_\varphi \subset C(-r,0;R^n)$ are open sets, $g: \mathcal{O}_x \times \mathcal{O}_u \times T \to R$, $f: \mathcal{O}: = \mathcal{O}_\varphi \times \mathcal{O}_u \times T \to R^n$, and $\Omega \subset \mathcal{O}_u$. (Here the indices in $\mathcal{O}_x$, $\mathcal{O}_u$ etc. refer to the variables chosen in these sets.)

In this problem, controls u in the following set of admissible controls are allowed:

$$\mathcal{U}_{ad}: = \{u \in L^\infty(T;R^m): u(t) \in \Omega \text{ a.e.}\}. \tag{1.1}$$

Observe that

$$\mathcal{U}_{ad} \subset \mathcal{O}_U: = \{u \in L^\infty(T;R^m): u(t) \in \mathcal{O}_u \text{ a.e.}\}.$$

We will prove a necessary optimality condition for pairs (x^o,u^o) which are optimal in the following sense.

Definition 1.2 Let $x^0 \in C(t_1-r,t_2;R^n)$ and $u^0 \in L^\infty(t_1,t_2;R^m)$ satisfy the constraints of Problem 1.1. The pair (x^0,u^0) is called a *strong local minimum* of Problem 1.1 if there is $\varepsilon > 0$ such that for all pairs (x,u) satisfying the constraints of Problem 1.1 with $|x-x^0| < \varepsilon$ the following inequality holds:

$$1/(t_2-t_1) \int_{t_1}^{t_2} g(x^0(t),u^0(t),t)dt \le 1/(t_2-t_1) \int_{t_1}^{t_2} g(x(t),u(t),t)dt \tag{1.2}$$

Strong local minima are global minima with respect to u.

Henceforth (x^0,u^0) will denote a strong local minimum of Problem 1.1 and $\varphi^0 := x^0_{t_1} \in \mathcal{O}_\varphi$.

Extend f,g,x^0, and u^0 in a periodic way to the whole real line. Then x^0,u^0 satisfy the system equation for a.a. $t \in R$.

Throughout this chapter, we assume, mostly without further mentioning, that the following hypotheses are satisfied.

Hypothesis 1.3 The functions $g(x,u,t)$ and $f(\varphi,u,t)$ are measurable in t and jointly continuous in (x,u) and (φ,u), respectively, and continuously Fréchet differentiable in x and φ, respectively.

Hypothesis 1.4 There exists a monotonically increasing function $q: R_+ \to R_+$ such that for all $x \in \mathcal{O}_x$, $\varphi \in \mathcal{O}_\varphi$, $\omega \in \Omega$ and a.a. $t \in T$

$$|g(x,\omega,t)| + |g_x(x,\omega,t)| \le q(|x|)$$

$$|f(\varphi,\omega,t)| + |\mathcal{D}_1 f(\varphi,\omega,t)| \le q(|\varphi|).$$

Hypothesis 1.5 For every $u \in \mathcal{U}_{ad}$ the initial value problem

$$x_{t_1} = \varphi^0, \quad \dot{x}(t) = f(x_t,u(t),t) \quad \text{a.a. } t \in T \tag{1.3}$$

has a solution $x(u,\varphi^0)$ on T and $x(u,\varphi^0)_t \in \mathcal{O}_\varphi$ is uniformly bounded for $u \in \mathcal{U}_{ad}$, say

$$|x(u,\varphi^0)_t| \le c_0, \quad \text{for all } t \in T.$$

For the linearized system, we assume that the fundamental extendability property (III.2.2) holds:

Hypothesis 1.6 For all x^1 and u^1 with $|x^1-x^0|_\infty$ and $|u^1-u^0|_2$ small enough there exists $k \in L^2(t_1-r,t_2;R)$ such that

$$\int_{t_1}^{t_2} |\mathcal{D}_1 f(x_t^1, u_1(t), t) x_t| dt \le |k|_2 |x|_2$$

for all $x \in C(t_1-r, t_2; R^n)$.

The Hypotheses 1.3 and 1.4 imply local existence and uniqueness for solutions of the initial value problem (1.3) with initial functions $x_{t_1} = \varphi$ in a neighborhood of φ^0 (cp. Section III.3). Hypothesis 1.5 is e.g. satisfied if for all $\varphi \in \mathcal{O}_\varphi$, all $\omega \in \Omega$ and a.a. $t \in T$

$$\varphi(0)^T f(\varphi, \omega, t) \le c(1+|\varphi|^2)$$

for some constant $c > 0$. This follows as Proposition III.3.1.

For simplicity assume that one can choose open sets $\mathcal{O}_\varphi, \mathcal{O}_U \supset \mathcal{U}_{ad}$, and $\mathcal{O}$ such that $(x^0, u^0, \varphi^0) \in \mathcal{O} \times \mathcal{O}_U \times \mathcal{O}_\varphi$ and the following map with values in $C(T;R^n)$ is well defined on $\mathcal{O} \times \mathcal{O}_U \times \mathcal{O}_\varphi$:

$$[F(x,u,\varphi)](t) := \varphi(0) + \int_{t_1}^{t} f(x_s, u(s), s) ds, \quad t \in T \tag{1.4}$$

where $x_{t_1} := \varphi$ (here and in the following we use for the restriction of x to T the same symbol x).

Then F is continuously Frêchet differentiable with respect to x and the equation

$$x = F(x, u, \varphi) \tag{1.5}$$

has, for every $(u,\varphi) \in \mathcal{O}_U \times \mathcal{O}_\varphi$, a unique solution

$$x = S(u, \varphi); \tag{1.6}$$

the solution map $S: \mathcal{O}_U \times \mathcal{O}_\varphi \to C(T;R^n)$ is continuous and continuously Frêchet differentiable with respect to the second argument; the derivative $x = \mathcal{D}_2 S(u^1, \varphi^1)\varphi$ is the unique solution of

$$x_{t_1} = \varphi, \dot{x}(t) = \mathcal{D}_1 f(x_t^1, u^1(t), t) x_t \quad \text{a.a. } t \in T \tag{1.7}$$

where $x^1 := S(u^1, \varphi^1)$.

All this follows as the results in Section III.3.

By choosing $\mathcal{O}_\varphi$ small enough, one gets a map

$$J: \mathcal{O}_U \times \mathcal{O}_\varphi \to R$$

$$J(u,\varphi) := 1/(t_2-t_1) \int_{t_1}^{t_2} g(S(u,\varphi)(t),u(t),t)dt \tag{1.8}$$

which is continuous and continuously Fréchet differentiable with respect to φ; the derivative satisfies

$$\mathcal{D}_2 J(u^1,\varphi^1)\varphi = 1/(t_2-t_1) \int_{t_1}^{t_2} g_x(x^1(s),u^1(s),s)x(s)ds, \tag{1.9}$$

where $x^1 := S(u^1,\varphi^1)$ and x is the unique solution of

$$x_{t_1} = \varphi, \dot{x}(t) = \mathcal{D}_1 f(x_t^1,u^1(t),t)x_t \quad \text{a.a. } t \in T. \tag{1.10}$$

Strong local minima enjoy the following property.

Lemma 1.7 The pair (x^0,u^o) is a strong local minimum of Problem 1.1 iff $x^0 = S(u^o,\varphi^0)$ and there are a neighborhood $\mathcal{O}_\varphi$ of φ^0 in $C(-r,0;R^n)$ and $\varepsilon > 0$ such that

$$J(u^o,\varphi^0) \le J(u,\varphi)$$

for all $u \in \mathcal{U}_{ad}$ and all $\varphi \in \mathcal{O}_\varphi$ with $S(u,\varphi)_{t_2} = \varphi$ and $|S(u,\varphi)-x^0) < \varepsilon$.

Proof: Clear by the preceeding. □

2. A Global Maximum Principle

We will prove the following two theorems.

Theorem 2.1 Let $t_2 \ge t_1+r$ and suppose that (x^0,u^o) is a strong local minimum of Problem 1.1 with Hypotheses 1.3 - 1.6 holding.

Then there exist $\lambda_0 \ge 0$ and a (t_2-t_1)-periodic solution y of the adjoint equation

$$\frac{d}{ds}\{y(s) + \int_s^{t_2+r} [\eta^T(\alpha,s-\alpha)-\eta^T(\alpha,t_2-\alpha)]y(\alpha)d\alpha\} \tag{2.1}$$
$$= -\lambda_0 g_x(x^0(s),u^o(s),s) \quad \text{a.a. } s \in T$$

such that $(\lambda_0,y^{t_2}) \neq (0,0)$ in $R\times NBV(0,r;R^n)$ and the following minimum condition holds:

$$\lambda_0 g(x_s^o,u^o(s),s) + y(s)^T f(x_s^o,u^o(s),s) \tag{2.2}$$
$$\le \lambda_0 g(x_s^o,\omega,s) + y(s)^T f(x_s^o,\omega,s)$$

for a.a. $s \le t_2$ and all $\omega \in \Omega$;

here η is given by the representation (cp. Lemma III.1.1)

$$\mathcal{D}_1 f(x_t^0, u^0(t), t)\varphi = \int_{-r}^{0} [d_s\eta(t,s)]\varphi(s), \quad \varphi \in C(-r,0;R^n).$$

The next theorem provides an additional condition for problems where the time interval is also chosen in an optimal way. For simplicity we deal only with the autonomous problem, where f and g are independent of t. Let (x^0, u^0, τ^0) satisfying the constraints of Problem 1.1 (with $t_1 = 0$, $t_2 = \tau^0 > r$) be optimal in the following sense:

For every triple (x,u,τ) satisfying the constraints of Problem 1.1 (with $t_1 = 0$, $t_2 = \tau$) and $|x-x^0|_\infty < \varepsilon$ and $|\tau-\tau^0| < \varepsilon$ for some $\varepsilon > 0$ the following inequality holds:

$$\frac{1}{\tau^0}\int_0^{\tau^0} g(x^0(s),u^0(s))ds \le \frac{1}{\tau}\int_0^{\tau} g(x(s),u(s))ds.$$

<u>Theorem 2.2</u> Let (x^0, u^0, τ^0) be optimal in the sense specified above for the autonomous version of Problem 1.1, suppose that the assumptions of Theorem 2.1 are satisfied and, additionally, that $\dot{x}_\tau^0$ is continuous and τ is a Lebesgue point of $u^0(\cdot)$.

Then the adjoint equation (2.1) and the maximum condition (2.2) are satisfied (with $t_1 = 0$, $t_2 = \tau^0$) and, additionally, the following transversality condition holds:

$$(F^*(\tau^0)y^{\tau^0})(\dot{x}^0)_{\tau^0} - \lambda_0 g(x^0(\tau^0),u^0(\tau^0)) = \lambda_0/\tau^0 \int_0^{\tau^0} g(x^0(s),u^0(s))ds; \quad (2.3)$$

here $F^*(\cdot)$ is the structural operator of the adjoint equation (2.1) (cp. (III.1.12)).

<u>Proof of Theorem 2.1:</u> First recall that the time interval $[t_1,t_2]$ is kept fixed here. Hence, by redefining g appropriately and later multiplying y by (t_2-t_1) we omit the factor $1/(t_2-t_1)$ before g throughout the proof.

By Lemma 1.7 there exist $\varepsilon > 0$ and a closed neighborhood $\mathcal{O}_\varphi$ of φ^0 in $C(-r,0;R^n)$ such that

$$J(u^0,\varphi^0) \le J(u,\varphi)$$

for all $(u,\varphi) \in V := \mathcal{U}_{ad} \times \mathcal{O}_\varphi$ with $|S(u,\varphi)-x^0| < \varepsilon$.

The following proof uses strong variations of u^0 of the form

$$u_{s,\omega}^{\rho}(t) := \begin{cases} u^0(t), & t_1 \le t < s-\rho \text{ and } s < t \le t_2 \\ \omega, & s-\rho \le t \le s; \end{cases}$$

here $\rho > 0$, $\omega \in \Omega$ and $s \in (t_1+\rho, t_2]$.

Where no confusion should possibly arise, the shorthand u^ρ for $u^\rho_{s,\omega}$ is used. Note that for ρ small enough $u^\rho \in \mathcal{U}_{ad}$, and let

$$x^\rho := S(u^\rho, \varphi^0).$$

By Hypothesis 1.5, x^ρ exists and $x^\rho - x^0$ is bounded, uniformly for $\rho > 0$. Let $\omega \in \Omega$ and choose $s \in T$ as a Lebesgue point of $f(x^0_t, u^0(t), t)$ and $f(x^0_t, \omega, t)$, $t \in T$.

First, limit properties for $\rho \to 0$ are analysed. Then Ekeland's Variational Principle is applied and the adjoint equation and the minimum condition are derived using the structure theory of Section III.1.

Lemma 2.3 There exists a sequence (ρ_i) tending to zero such that for all $t \in T$

$$\bar{x}(t) := \lim 1/\rho_i [x^{\rho_i}(t) - x^0(t)]$$

exists, vanishes on $[t_1, s)$ and coincides for $t \in [s, t_2]$ with the unique solution of the initial value problem

$$x(t) = 0, \quad t < s, \quad x(s) = f(x^0_s, \omega, s) - f(x^0_s, u^0(s), s),$$

$$\dot{x}(t) = \mathcal{D}_1 f(x^0_t, u^0(t), t) x_t, \quad \text{a.a. } t \in [s, t_2];$$

furthermore

$$\lim |\bar{x} - 1/\rho_i [x^{\rho_i} - x^0]|_{L_2(T;R^n)} = 0.$$

Proof: Note first that $x(s)$ is well-defined by choice of s; furthermore the initial value problem above has a unique solution.

The proof of the lemma will be subdivided into three steps.

Step 1 $1/\rho \sup_{t \in T} |x^\rho(t) - x^0(t)|$ is uniformly bounded for $\rho \to 0$.

Proof: Recall that for $t \le s-\rho$ one has $x^\rho(t) = x^0(t)$. For $t > s-\rho$

$$1/\rho |x^\rho(t) - x^0(t)|$$

$$\le 1/\rho \int_{s-\rho}^{t} |f(x^\rho_\sigma, u^\rho(\sigma), \sigma) - f(x^0_\sigma, u^0(\sigma), \sigma)| d\sigma$$

$$\le 1/\rho \int_{s-\rho}^{s} |f(x^\rho_\sigma, \omega, \sigma) - f(x^0_\sigma, u^0(\sigma), \sigma)| d\sigma$$

$$+ 1/\rho \int_{s}^{t} \max_{\alpha \in [0,1]} |\mathcal{D}_1 f(x^0_\sigma + \alpha(x^\rho_\sigma - x^0_\sigma), u^0(\sigma), \sigma)| \, |x^\rho_\sigma - x^0_\sigma| d\sigma$$

by definition of ρ and the mean value theorem;

$$\leq 2q(c_0) + 1/\rho \int_s^t q(c_0)|x_\sigma^\rho - x_\sigma^0| d\sigma$$

using Hypotheses 1.4 and 1.5. Invoke Gronwall's inequality (see e.g. Hale [1977,Lemma 3.1,p.15]) in order to see that this implies

$$1/\rho|x_t^\rho - x_t^0| \leq c_1 + c_2 q(c_0)(t_2 - t_1) \tag{2.4}$$

for constants $c_1, c_2 > 0$.

Step 2 $\lim_{\rho\to 0} 1/\rho[x^\rho(t) - x^0(t)] = 0, \quad t \in [t_1, s)$ and

$\lim_{\rho\to 0} 1/\rho[x^\rho(s) - x^0(s)] = f(x_s^0,\omega,s) - f(x_s^0,u^0(s),s).$

Proof: The first assertion is clear by the definitions, and

$$1/\rho[x^\rho(s) - x^0(s)]$$

$$= \int_{s-\rho}^{s} 1/\rho[f(x_\sigma^\rho,\omega,\sigma) - f(x_\sigma^0,\omega,\sigma)]d\sigma$$

$$+ 1/\rho \int_{s-\rho}^{s} [f(x_\sigma^0,\omega,\sigma) - f(x_\sigma^0,u^0(\sigma),\sigma)]d\sigma.$$

For $\rho \to 0$ the integrand in the first term remains bounded by the mean value theorem and boundedness of $1/\rho[x_\sigma^\rho - x_\sigma^0]$; hence the integral converges to zero. The second integral converges to

$$f(x_s^0,\omega,s) - f(x_s^0,u^0(s),s)$$

by choice of s.

Step 3 For $s \leq t \leq t_2$, a subsequence of $(1/\rho[x^\rho(t) - x^0(t)])$ converges uniformly to the solution of the initial value problem in the lemma.

Proof: We know that $1/\rho[x^\rho(t) - x^0(t)]$ is uniformly bounded, say by M; furthermore it is also equicontinuous, since for $s \leq \tau \leq t \leq t_2$

$$|1/\rho[x^\rho(t)-x^0(t)] - 1/\rho[x^\rho(\tau) - x^0(\tau)]|$$

$$\leq \int_\tau^t \sup_{\alpha\in[0,1]} |D_1 f(x_\sigma^\rho + \alpha(x_\sigma^0 - x_\sigma^\rho),u^0(\sigma),\sigma)|\,1/\rho|x_\sigma^\rho - x_\sigma^0| d\sigma$$

$$\leq q(c_0)M(t-\tau).$$

Hence by the Arzêla-Ascoli Theorem a subsequence converges uniformly to a continuous function, say $\bar{x}(\cdot)$, on $[s,t_2]$. We know

$$\bar{x}(s) = f(x_s^0,\omega,s) - f(x_s^0,u^0(s),s)$$

and define

$$\bar{x}(t) := 0 \quad \text{for} \quad t < s.$$

For every $t \in [s,t_2]$

$$\int_s^t \mathcal{D}_1 f(x_\sigma^o,u^o(\sigma),\sigma)1/\rho[x_\sigma^\rho - x_\sigma^o]d\sigma \to \int_s^t \mathcal{D}_1 f(x_\sigma^o,u^o(\sigma),\sigma)\bar{x}_\sigma d\sigma$$

and

$$|1/\rho[x^\rho(t)-x^0(t)] - 1/\rho[x^\rho(s)-x^0(s)] - \int_s^t \mathcal{D}_1 f(x_\sigma^o,u^o(\sigma),\sigma)1/\rho[x_\sigma^\rho - x_\sigma^o]d\sigma|$$

$$= |\int_s^t 1/\rho\{f(x_\sigma^\rho,u^o(\sigma),\sigma) - f(x_\sigma^o,u^o(\sigma),\sigma) - \mathcal{D}_1 f(x_\sigma^o,u^o(\sigma),\sigma)[x_\sigma^\rho - x_\sigma^o]\}\, d\sigma|$$

$$\leq \int_s^t \sup_{\alpha\in[0,1]} |\mathcal{D}_1 f(x_\sigma^\rho + \alpha(x_\sigma^o - x_\sigma^\rho),u^o(\sigma),\sigma) - \mathcal{D}_1 f(x_\sigma^o,u^o(\sigma),\sigma)|M d\sigma.$$

The right hand side tends to zero by the Dominated Convergence Theorem. This concludes the proof of Step 3, and the lemma is proven, noting that the last assertion again follows by the Dominated Convergence Theorem.

□

Lemma 2.4 For a subsequence (ρ_i) tending to zero

$$\lim 1/\rho_i[J(u^{\rho_i},\varphi^0) - J(u^o,\varphi^0)] = [g(x^0(s),\omega,s) - g(x^0(s),u^o(s),s)]$$

$$+ \int_s^{t_1} g_x(x^0(\sigma),u^o(\sigma),\sigma)\{\Phi(\sigma,s)X_o[f(x_s^o,\omega,s) - f(x_s^o,u^o(s),s)]\}(0)d\sigma;$$

here $\Phi(\sigma,s)$ is the evolution operator associated with (1.10).

Proof: By the definitions

$$1/\rho[J(u^\rho,\varphi^0) - J(u^o,\varphi^0)]$$

$$= 1/\rho \int_{s-\rho}^{t_2} [g(x^\rho(s),u^\rho(s),s) - g(x^0(s),u^o(s),s)]ds$$

$$= 1/\rho\{\int_{s-\rho}^{s} [g(x^\rho(s),\omega,s) - g(x^0(s),\omega,s)]ds$$

$$+ \int_{s-\rho}^{s} [g(x^0(s),\omega,s) - g(x^0(s),u^o(s),s)]ds$$

$$+ \int_s^{t_2} [g(x^\rho(s),u^o(s),s) - g(x^0(s),u^o(s),s)]ds\}.$$

Arguing as in the proof of the preceeding lemma, one obtains the assertion, noting that for $\sigma \geq s$

$$x_\sigma = \Phi(\sigma,s)X_o[f(x_s^o,\omega,s) - f(x_s^o,u^o(s),s)]$$

where x_σ is the solution of the initial value problem formulated in Lemma 2.3.

□

For an application of Corollary II.3.7, we prepare the following lemma.

Lemma 2.5 For $u,v \in \mathcal{U}_{ad}$ define

$$d(u,v) := \text{meas}\{t \in T: u(t) \neq v(t)\}.$$

Then d defines a metric on $\mathcal{U}_{ad}$ and $\mathcal{U}_{ad}$ is complete in this metric. Hence also $V = \mathcal{U}_{ad} \times \mathcal{O}_\varphi$ is a complete metric space where $\mathcal{O}_\varphi$ is endowed with the metric induced by the uniform norm.

Proof: See Ekeland [1979,p.454] for the proof that d defines a complete metric. □

Henceforth, V will always be considered in the metric topology defined above.

Now take a sequence $\delta_n \to 0$, $\delta_n > 0$ and define functionals $F_n: V \to R$ by

$$F_n(u,\varphi) := [\,|x(u,\varphi)_{t_2} - \varphi|_M + |J(u,\varphi) - (m-\delta_n)|^2]^{1/2} \tag{2.5}$$

where $m := J(u^o,\varphi^o)$, and the norm in the first summand is taken in $M := M^2(-r,0;R^n)$.

Lemma 2.6 For every $n \in N$, the functional F_n is continuous on $\mathcal{U}_{ad} \times \mathcal{O}_\varphi$ and

$$F_n(u,\varphi) > 0 \quad \text{for all} \quad (u,\varphi) \in V,$$

where V is a closed (metric) neighborhood of (u^o,φ^o).

Proof: Continuity follows by Hypothesis 1.5 and the Arzêla-Ascoli Theorem. Furthermore, the assumptions

$$F_n(u,\varphi) = 0 \quad \text{and} \quad J(u,\varphi) = m - \delta_n < m$$

contradict optimality of (u^o,φ^o). □

Naturally, the sequence (ε_n) with $\varepsilon_n := F_n(u^o,\varphi^o)$ converges to zero and

$$F_n(u^o,\varphi^o) \leq \inf\{F_n(u,\varphi): (u,\varphi) \in V\} + \varepsilon_n.$$

Thus by Corollary II.3.7, applied to F_n, there exist $(u^n,\varphi^n) \in V$ with the following properties:

$$0 \leq F_n(u^n,\varphi^n) \leq F_n(u^o,\varphi^o) = \varepsilon_n \tag{2.6}$$

$$d(u^o,u^n) \le \varepsilon_n^{1/2}, \quad |\varphi^0-\varphi^n| \le \varepsilon_n^{1/2} \tag{2.7}$$

$$F_n(u,\varphi) \ge F_n(u^n,\varphi^n) - \varepsilon_n^{1/2}[d(u,u^n) + |\varphi-\varphi^n|] \tag{2.8}$$

for all $(u,\varphi) \in V$.

Let $s \in T$, $\omega \in \Omega$ and $\rho > 0$, small enough. We shall use the relations above for $(u,\varphi) = (u^{n,\rho}_{s,\omega},\varphi^n)$ and for $(u,\varphi) = (u^n,\varphi^0+\rho\psi)$.

Note that the functional on $M\times R$

$$(z,\xi) \to (|z|^2 + \xi^2)^{1/2}$$

has the Fréchet derivative in direction (y,η) at (z,ξ)

$$(|z|^2 + \xi^2)^{-1/2}(z^*y + \xi\eta) \tag{2.9}$$

if $|z|^2 > 0$ and $\xi^2 > 0$; here $z^* \in M^*$ is the functional corresponding to z.

We obtain from (2.8) for $(u,\varphi) \neq (u^n,\varphi^n)$ the following important inequality:

$$-\varepsilon_n^{1/2} \le \{F_n(u,\varphi) - F_n(u^n,\varphi^n)\}/[d(u,u^n) + |\varphi-\varphi^n|] \tag{2.10}$$

$$= \{[x(u,\varphi)_{t_2}-\varphi^n|_M^2 + |J(u,\varphi)-(m-\delta_n)|^2]^{1/2}$$

$$- [|x(u^n,\varphi^n)_{t_2}-\varphi^n|_M^2 + |J(u^n,\varphi^n)-(m-\delta_n)|^2]^{1/2}\}/[d(u,u^n) + |\varphi-\varphi^n|].$$

Let $x^n: = x(u^n,\varphi^n)$ and define $z^{n*} \in M^*$ and $\lambda^n \in R$ by

$$z^{n*}: = [|x^n_{t_2}-\varphi^n|_M^2 + |J(u^n,\varphi^n)-(m-\delta_n)|^2]^{-1/2}[x^{n*}_{t_2}-\varphi^{n*}] \tag{2.11}$$

$$\lambda^n: = [|x^n_{t_2}-\varphi^n|_M^2 + |J(u^n,\varphi^n)-(m-\delta_n)|^2]^{-1/2}[J(u^n,\varphi^n)-(m-\delta_n)] \tag{2.12}$$

Let $\Phi^n(t,s)$ be the family of evolution operators associated with the linearized equations $(n = 0,1,\ldots)$

$$\dot{x}(t) = \mathcal{D}_1 f(x^n_t,u^n(t),t)x_t, \quad \text{a.a. } t \in T; \tag{2.13}$$

for Φ^o we also write Φ.

By Hypothesis 1.6, the operators Φ^n have continuous extensions Φ^n_M to M.

Abbreviate

$$T_n := \Phi_M(t_2,t_1) - Id. \tag{2.14}$$

Now take $(u,\varphi) = (u^n,\varphi^n + \rho\psi)$ in (2.10). The chain rule and (2.9) imply that F_n is continuously Fréchet differentiable with respect to φ. Thus in the limit for $\rho \to 0$, we get from (2.10)

$$-\varepsilon_n^{1/2} \le z^{n*}T^n\iota\psi + \lambda^n \int_{t_1}^{t_2} g_x(x^n(\sigma),u^n(\sigma),\sigma)[\Phi_M^n(\sigma,t_1)\iota\psi](0)d\sigma.$$

Since ψ is arbitrary in $C(-r,0;R^n)$ this implies

$$(T^{n*}z^{n*})\psi = -\lambda^n \int_{t_1}^{t_2} g_x(x^n(\sigma),u^n(\sigma),\sigma)[\Phi_M^n(\sigma,t_1)\psi](0)d\sigma. \tag{2.15}$$

for all $\psi \in M$.

Now we consider limits for $n \to \infty$. By (2.7), $|\varphi^n-\varphi^0| \to 0$, hence, by continuity, resp. continuous Fréchet differentiability, we get

$|x^n-x^0|_\infty \to 0$, $|\Phi^n(\sigma,t_1)-\Phi(\sigma,t_1)| \to 0$ for all $\sigma \in T$,

$|T^n-T^o| \to 0$ (first on C, then on M) and

$|\Phi_M^n(\sigma,s)|$ is uniformly bounded for $t_2 \ge \sigma \ge s \ge t_1$.

Recall that by definition $|\lambda^n| \le 1$ and $|z^{n*}|_{M*} \le 1$. We have to exclude that both sequences (λ^n) and (z^{n*}) converge to zero. Suppose first that the Fredholm operator $\Phi_M(t_2,t_1)$ - Id is surjective, hence an isomorphism of M. This implies that, for sufficiently large n, also $T^n = \Phi_M^n(t_2,t_1)$ - Id and hence T^{n*} are isomorphisms (cp. e.g. Dunford/Schwartz [1967, Lemma VII.6.1]).

Suppose that there exists a subsequence of (λ^n) again denoted by (λ^n) converging to zero. This yields (look at (2.11), (2.12))

$$|z^{n*}|_{M*} \to 1.$$

This implies existence of $(\psi^n) \in M$ with

$$|\psi^n| \le 1 \quad \text{and} \quad (T^{n*}z^{n*})\psi^n \ge 1/2.$$

But from (2.15) we obtain

$$|(T^{n*}z^{n*})\psi^n| \le |\lambda^n|c_0,$$

where c_0 is a constant independent of n.

This is a contradiction, since $\lambda^n \to 0$. Thus zero cannot be a clusterpoint of (λ^n).

Let λ_0 be a clusterpoint of (λ^n) and z^{o*} be a weak* clusterpoint of (z^{n*}), which exists since $|z^n| \le 1$.

Now (2.15) and (III.2.8) imply for all $\varphi \in C(-r,0;R^n)$

$$[T^{o*}z^{o*}]\iota\varphi = [\iota^*(\Phi_M(t_2,t_1)-Id)^*z^{o*}]\varphi = [(\Phi(t_2,t_1)-Id)^*\iota^*z^{o*}]\varphi$$

$$= -\lambda_0 \int_{t_1}^{t_2} g_x(x^0(\sigma),u^o(\sigma),\sigma)[\Phi(\sigma,t_1)\varphi](0)d\sigma.$$

Define

$$y^*: = \iota^*z^{o*} \in C(-r,0;R^n)^* = NBV(-r,0;R^n).$$

(recall (III.2.10) and the ensuing discussion).

Then

$$[(\Phi(t_2,t_1)-Id)^*y^*]\varphi = -\lambda_0 \int_{t_1}^{t_2} g_x(x^0(\sigma),u^o(\sigma),\sigma)[\Phi(\sigma,t_1)\varphi](0)d\sigma \quad (2.16)$$

for all $\varphi \in C(-r,0;R^n)$.

If $\Phi(t_2,t_1) - Id$ is not an isomorphism, there exists

$0 \neq y^* \in NBV(-r,0;R^n)$ such that, with $\lambda_0 = 0$, again

(2.16) holds.

We note, also for reference in later chapters, the following consequence.

Lemma 2.7 Suppose that $t_2 \ge t_1+r$.
Then equation (2.16) implies that there exists $\psi \in NBV(0,r;R^n)$ with

$$y^* = F^*(t_2)\psi. \quad (2.17)$$

Here $F(t)$ and $\Phi(t,s)$ are the structural operators and the family of evolution operators, respectively, of the linear retarded equation

$$\dot{x}(t) = \mathcal{D}_1 f(x_t^o,u^o(t),t)x_t, \quad \text{a.a. } t \in R.$$

Proof: Equation (2.16) implies that for all $\varphi \in C(-r,0;R^n)$

$$y^*\psi = [\Phi^*(t_2,t_1)y^*]\varphi-\lambda_0 \int_{t_1}^{t_2} g_x(x^0(t),u^o(t),t)[\Phi(t,t_1)\varphi](0)dt.$$

But for $t_2 \ge t_1+r$

$$\Phi(t_2,t_1)y^* = [\Phi(t_2,t_1+r)\Phi(t_1+r,t_1)]^*y^*$$

$$= F^*(t_1)G^*(t_1)\Phi^*(t_2,t_1+r)y^* \in \text{Im } F^*(t_2),$$

since, by periodicity, $F(t_1) = F(t_2)$.

Furthermore observe that one can write

$$\int_{t_1}^{t_2} g_x(x^0(t),u^o(t),t)[\Phi(t,t_1)\varphi](0)dt$$
$$= \gamma_o\Phi(t_1+r,t_1)\varphi + \int_{t_1+r}^{t_2} \gamma(t)\Phi(t,t_1+r)\Phi(t_1+r,t_1)\varphi dt$$

for elements $\gamma_o,\gamma(t) \in C(-r,0;R^n)^* = NBV(-r,0;R^n)$. Hence by Proposition III.1.4

$$= [F^*(t_1)G^*(t_1)\gamma_o]\varphi + F^*(t_1) \int_{t_1+r}^{t_2} G^*(t_1)\Phi^*(t,t_1+r)\gamma(t)]\varphi dt.$$

Thus the lemma follows. □

Using (2.17) in (2.16) we obtain for all $\psi \in C(-r,0;R^n)$

$$0 = \Phi^*(t_2,t_1)F^*(t_2)\psi - F^*(t_2)\psi \tag{2.18}$$
$$+ \lambda_0 \int_{t_1}^{t_2} g_x(x^0(\sigma),u^o(\sigma),\sigma)[\Phi(\sigma,t_1)\psi](0)d\sigma.$$

By equations (III.1.24) and (III.1.29), it follows that

$$0 = F^*(t_1)\psi^*(t_2,t_1)\psi - F^*(t_2)\psi$$
$$+ \lambda_0 \int_{t_1}^{t_2} \int_{-r}^{o} [d_s(F^*(\sigma)Y_o g_x(x^0(\sigma),u^o(\sigma),\sigma))(s)]^T[\Phi(\sigma,t_1)\psi](s)d\sigma$$
$$= F^*(t_1)\Psi^*(t_2,t_1)\psi - F^*(t_2)\psi$$
$$+ \lambda_0 F^*(t_1) \int_{t_1}^{t_2} [\Psi^*(\sigma,t_1)(Y_o g_x(x^0(\sigma),u^o(\sigma),\sigma))]\psi d\sigma.$$

Define the *adjoint equation* in $NBV(0,r;R^n)$ as

$$y^s = \Psi^*(t_2,s)\psi + \int_s^{t_2} \Psi^*(t,s)Y_o\lambda_o(x^0(t),u^o(t),t)dt, \quad s \in T \tag{2.19}$$

where ψ is given by (2.17); this equation is equivalent to (2.1) with $y^{t_2} = \psi$.

By (2.18)

$$F^*(t_2)\ \psi = F^*(t_2)y^{t_2} = F^*(t_1)y^{t_1}.$$

If necessary, we redefine $y^{t_2} := y^{t_1}$. Thus the solution $y(\cdot)$ of (2.19) is a (t_2-t_1)-periodic solution of (2.1). This establishes the adjoint equation.

Now take $(u,\varphi) = (u^{n,\rho},\varphi^n)$ in (2.10) in order to derive the minimum condition. Note that $d(u^{n,\rho},u^n) = \rho$, and hence

$$-\varepsilon_n^{1/2} \le [F_n(u^{n,\rho},\varphi^n) - F_n(u^n,\varphi^n)]/\rho \tag{2.21}$$
$$= \{[|x(u^{n,\rho},\varphi^n)_{t_2} - \varphi^n|_M^2 + |J(u^{n,\rho},\varphi^n)-(m-\delta_n)|^2]^{1/2}$$
$$- [|x(u^n,\varphi^n)_{t_2} - \varphi^n|_M^2 + |J(u^n,\varphi^n)-(m-\delta_n)|^2]^{1/2}\}/\rho.$$

Take s as a Lebesgue point of $f(x_s^n,\omega,s) - f(x_s^n,u^n(s),s)$, $s = 0,1,\ldots$ For $\rho \to 0$ one gets

$$-\varepsilon_n^{1/2} \le z^{n*}\iota\Phi^n(t_2,s)X_0[f(x_s^n,\omega,s) - f(x_s^n,u^n(s),s)]$$
$$+ \lambda^n \int_s^{t_2} g_x(x^n(\sigma),u^n(\sigma),\sigma)[\Phi^n(\sigma,s)X_0(f(x_s^n,\omega,s) - f(x_s^n,u^n(s),s))](0)d\sigma$$
$$+ \lambda^n[g(x^n(s),\omega,s) - g(x^n(s),u^n(s),s)],$$

and in the limit for $n \to \infty$

$$0 \le z^{0*}\iota\Phi(t_2,s)X_0[f(x_s^0,\omega,s) - f(x_s^0,u^0(s),s)]$$
$$+ \lambda_0 \int_s^{t_2} g_x(x^0(\sigma),u^0(\sigma),s)[\Phi(\sigma,s)X_0(f(x_s^0,\omega,s) - f(x_s^0,u^0(s),s))](0)d\sigma$$
$$+ \lambda_0[g(x^0(s),\omega,s) - g(x^0(s),u^0(s),s)].$$

Using the adjoint equation, we can rewrite this in the following way:

$$0 \le y^*[\Phi(t_2,s)X_0(f(x_s^0,\omega,s) - f(x_s^0,u^0(s),s))]$$
$$+ \lambda_0 \int_s^{t_2} g_x(x^0(\sigma),u^0(\sigma),\sigma)[\Phi(\sigma,s)X_0(f(x_s^0,\omega,s) - f(x_s^0,u^0(s),s))](0)d\sigma$$
$$+ \lambda_0[g(x_s^0,\omega,s) - g(x_s^0,u^0(s),s)]$$
$$= [F^*(t_2)\Psi^*(t_2,s)\psi][X_0(f(x_s^0,\omega,s) - f(x_s^0,u^0(s),s))]$$
$$+ \lambda_0 \int_s^{t_2}[F^*(t)\Psi^*(\sigma,s)Y_0 g_x(x_\sigma^0,u^0(\sigma),\sigma)][X_0(f(x_s^0,\omega,s)-f(x_s^0,u^0(s),s))]d\sigma$$
$$+ \lambda_0[g(x_s^0,\omega,s) - g(x_s^0,u^0(s),s)]$$
$$= y(s)^T[f(x_s^0,\omega,s) - f(x_s^0,u^0(s),s)] + \lambda_0[g(x_s^0,\omega,s) - g(x_s^0,u^0(s),s)].$$

This concludes the proof of Theorem 2.1. □

Proof of Theorem 2.2: The proof is analogous to the one given for Theorem 2.1 with the following modifications:

For a sequence $\delta_n \to 0$, $\delta_n > 0$ define

$$F_n(u,\varphi,\tau) := [|x(u,\varphi)_\tau - \varphi|_M^2 + |J(u,\varphi,\tau)-(m-\delta_n)|^2]^{1/2}$$

where $J(u,\varphi,\tau) := \frac{1}{\tau}\int_0^\tau g(S(u,\varphi)(t),u(t),t)dt$.

Application of Ekeland's Variational Principle yields the existence of (u^n,φ^n,τ^n) satisfying the analogs of (2.6), (2.7) and $|\tau^n-\tau^0| \le \varepsilon_n^{1/2}$,

$$F_n(u,\varphi,\tau) \ge F_n(u^n,\varphi^n,\tau^n) - \varepsilon_n^{1/2}[d(u,u^n) + |\varphi-\varphi^n| + |\tau-\tau^n|].$$

This implies (2.1) and (2.2) as above. However, one can in addition consider the derivative with respect to τ:

One obtains, taking $(u,\varphi,\tau) = (u^n,\varphi^n,\tau^n+\rho)$, in the limit for $\rho \to 0$

$$0 = z^{n*}\begin{pmatrix} \dot{x}(u^n,\varphi^n)(\tau^n) \\ \dot{x}(u^n,\varphi^n)_{\tau_n} \end{pmatrix} + \frac{d}{d\tau}\Big|_{\tau=\tau^n} \frac{\lambda^n}{\tau}\int_0^\tau g(x(u^n,\varphi^n)(s),u^n(s))ds$$

where we have used that the map

$$t \to x(u^n,\varphi^n)_t \quad : \quad R \to L^2(-r,0;R^n)$$

is differentiable: This follows from the fact that

$$\frac{1}{\varepsilon}[x(t+\varepsilon+s)-x(t+s)] = \frac{1}{\varepsilon}\int_t^{t+\varepsilon} \dot{x}(s+\tau)d\tau \to \dot{x}_t(s) \quad \text{in} \quad L_2(-r,0;R^n).$$

Choosing u^n such that τ^n is a Lebesgue point one finds

$$\frac{d}{d\tau}\Big|_{\tau=\tau^n} \frac{1}{\tau}\int_0^\tau g(x(u^n,\varphi^n)(s),u^n(s))ds$$
$$= -\frac{\lambda^n}{(\tau^n)^2}\int_0^{\tau^n} g(x(u^n,\varphi^n)(s),u^n(s))ds$$
$$+ \frac{\lambda^n}{\tau^n} g(x(u^n,\varphi^n)(\tau^n),u^n(\tau^n)).$$

Since $z^{n*} \to z^{o*}$ weakly* in M^* and

$$\begin{pmatrix} \dot{x}(u^n,\varphi^n)(\tau^n) \\ \dot{x}(u^n,\varphi^n)_{\tau^n} \end{pmatrix} \to \begin{pmatrix} \dot{x}(u^o,\varphi^0)(\tau^o) \\ \dot{x}(u^o,\varphi^0)_{\tau^o} \end{pmatrix} \quad \text{in} \quad M,$$

one obtains for $n \to \infty$

$$0 = z^{0*}\begin{pmatrix} \dot{x}(u^0,\varphi^0)(\tau^0) \\ \\ \dot{x}(u^0,\varphi^0)_{\tau^0} \end{pmatrix} - \frac{\lambda_0}{(\tau^0)^2}\int_0^{\tau^0} g(x^0(s),u^0(s))ds$$

$$+ \frac{\lambda_0}{\tau^0} g(x^0(\tau^0),u^0(\tau^0)).$$

Since $(\dot{x}^0)_{\tau^0}$ is continuous, the first summand equals

$$z^{0*}\iota(\dot{x}^0)_{\tau^0}$$

$$= (\iota^* z^{0*})(\dot{x}^0)_{\tau^0}$$

$$= y^*(\dot{x}^0)_{\tau^0}$$

$$= F^*(\tau^0)y^{\tau^0}(\dot{x}^0)_{\tau^0} .$$

Hence Theorem 2.2 follows. □

Remark 2.8 The idea for the proof of Theorem 2.1 goes back to work of Ekeland [1974] and Clarke [1976]. Ekeland [1979] gives a good account of the main ideas. The new ingredients in the proof above are duality theory of functional differential equations and use of the fact that $\Phi(t_2,t_1)$ - Id is Fredholm in order to get convergence of $\{z^n\}$ in an infinite dimensional space. See Fattorini [1985,1987] for different arguments in different infinite dimensional problems.

Remark 2.9 Sometimes it is convenient to rewrite the minimum condition using the Pontryagin function H defined as

$$H(\varphi,\omega,\lambda_0,y,t) := \lambda_0 g(\varphi(0),\omega,t) + y^T f(\varphi,\omega,t)$$

$$H: C(-r,0;R^n) \times R^m \times R \times R^n \times T \to R.$$

Then (2.2) is equivalent to

$$H(x_t^0,u^0(t),\lambda_0,y(t),t) = \min_{\omega\in\Omega} H(x_t^0,\omega,\lambda_0,y(t),t)$$

for a.a. $t \in T$.

Remark 2.10 If H is differentiable with respect to ω, condition (2.2) implies the following *local minimum condition*:

$$D_2H(x_t^0,u^0(t),\lambda_0,y(t),t)(\omega-u^0(t)) \geq 0$$

for a.a. $t \in T$ and all $\omega \in \Omega$.

Such a local version of the maximum principle for periodic control of functional differential equations was proven in Colonius [1986b]; a global version based on a generalization of Uhl's Theorem appears in Li [1985], Li/Chow [1987]. In these papers, extendability of the linearized system to the product space is not assumed, however, an a-priori assumption on the value function is made and the role of the adjoint equation is not clarified.

The transversality condition given in the latter reference is not valid, since the derivative with respect to τ is not performed correctly. Furthermore the transversality condition (2.3) above reduces to the one given by Gilbert [1977] for ordinary differential equations, while the one asserted by Li/Chow does not.

The first to treat optimal periodic control problems for functional differential equations were Sincic/Bailey [1978]. They gave a (formal) proof of a local maximum principle.

An easy consequence of the global maximum principle is the following *"Legendre-Clebsch Condition"*.

Corollary 2.11 Suppose that H is twice differentiable with respect to ω. Then condition (2.2) implies

$$\mathcal{D}_2\mathcal{D}_2 H(x_t^o,u^o(t),\lambda_o,y(t),t)(\omega-u^o(t),\omega-u^o(t)) \geq 0$$

for a.a. $t \in T$ and all $\omega \in \Omega$.

Remark 2.12 Suppose that the system is described by a delay equation with a single constant delay $r > 0$

$$\dot{x}(t) = f(x(t),x(t-r),u(t),t) \quad \text{a.a.} \quad t \in T,$$

where $f\colon R^n \times R^n \times R^m \times T \to R^n$.

In this case, the adjoint equation (2.1) reduces to

$$\begin{aligned}\dot{y}(s) = &- \mathcal{D}_1 f(x^o(s),x^o(s+r),u^o(s),s)^T y(s)\\ &- \mathcal{D}_2 f(x^o(s+r),x^o(s),u^o(s+r),s+r)^T y(s+r)\\ &+ \lambda_o g_x(x^o(s),u^o(s),s).\end{aligned}$$

Remark 2.13 In this chapter, we did not try to consider the most general problem. Presumably, also problems with state constraints might be included. For a treatment of semilinear equations in Hilbert space see Colonius [1987].

CHAPTER V

WEAK LOCAL MINIMA

In this chapter, first and second order necessary optimality conditions are proven using the Banach space methods of sections II.1 and II.2. While the first order conditions obtained in this way are weaker than those of Chapter IV (local versus global maximum principle), the second order conditions are new. They are our main justification for the use of these methods.

In Chapter VII second order conditions will turn out to be crucial in order to analyse properness. Furthermore, state constraints and more general boundary conditions are readily included in the treatment of the present chapter; in particular problems with fixed boundary values are discussed briefly in section 2 dealing with first order conditions.

The time interval is kept fixed throughout.
The main results of this chapter are Theorem 2.4, Theorem 2.9, Theorem 3.7 and Theorem 3.8.

1. Problem Formulation

We consider the following optimal control problem for functional differential equations:

Problem 1.1 Minimize $\int_{t_1}^{t_2} g(x(s),u(s),s)ds$

s.t. $\dot{x}(t) = f(x_t,u(t),t)$ a.a. $t \in T := [t_1,t_2]$

$h(x(t),t) \in R^{\ell}_{-}$ a. $t \in T$

$u(t) \in \Omega(t)$ a.a. $t \in T$

$p(x_{t_1},x_{t_2}) = 0$

here $g: \mathcal{O}_x \times \mathcal{O}_u \times T \to R$, $f: \mathcal{O}_\varphi \times \mathcal{O}_u \times T \to R^n$, $h: \mathcal{O}_x \times T \to R^{\ell}$, $p: \mathcal{O}_\varphi \times \mathcal{O}_\varphi \to Y$, with $\mathcal{O}_x \subset R^n$, $\mathcal{O}_u \subset R^m$, $\mathcal{O}_\varphi \subset C(-r,0;R^n)$ open, Y a Banach space, and $\Omega(t) \subset \mathcal{O}_u$ closed and convex.

Again we only allow essentially bounded controls u in

$$\mathcal{U}_{ad} := \{u \in L^\infty(T;R^m): u(t) \in \Omega(t) \text{ a.e.}\}. \tag{1.1}$$

Of special interest for us is the periodic problem:

<u>Problem 1.2</u> Minimize $1/(t_2-t_1) \int_{t_1}^{t_2} g(x(s),u(s),s)ds$

s.t. $\dot{x}(t) = f(x_t,u(t),t)$ a.a. $t \in T := [t_1,t_2]$

$h(x(t),t) \in R^\ell_-$ a. $t \in T$

$u(t) \in \Omega(t)$ a.a. $t \in T$

$x_{t_1} = x_{t_2}$

where f,g,h and $\Omega(\cdot)$ are as in Problem 1.1 and $u \in \mathcal{U}_{ad}$.

<u>Remark 1.3</u> Problem 1.2 is a special case of Problem 1.1, defining $p: C(-r,0;R^n)\times C(-r,0;R^n) \to Y := C(-r,0;R^n)$ by $p(\varphi,\psi) := \varphi-\psi$.

<u>Remark 1.4</u> After Remark 2.13 we will briefly discuss the fixed boundary value problem, where p in Problem 1.1 is specified as

$$p(\varphi,\psi) := (\varphi-\varphi^0,\psi-\psi^0)$$

and $\varphi^0,\psi^0 \in C(-r,0;R^n)$ are fixed (the appropriate choice of Y is discussed there).

We will prove necessary optimality conditions for pairs (x^0,u^0) which are optimal in the following sense.

<u>Definition 1.5</u> A pair $(x^0,u^0) \in C(t_1-r,t_2)\times L^\infty(t_1,t_2;R^m)$ satisfying the constraints of Problem 1.1 is called a *weak local minimum,* if for some $\varepsilon > 0$ and for all such pairs (x,u) with

$$|x-x^0|_\infty < \varepsilon \quad \text{and} \quad |u-u^0|_\infty < \varepsilon$$

the inequality

$$\int_{t_1}^{t_2} g(x^0(t),u^0(t),t)dt \le \int_{t_1}^{t_2} g(x(t),u(t),t)dt$$

holds.

Henceforth we assume that a weak local optimal solution (x^0,u^0) of Problem 1.1 is given. Define $\varphi^0 := x^0_{t_1} \in \mathcal{O}_\varphi \subset C(-r,0;R^n)$.

Every strong local minimum (cp. Definition IV.1.2) is a weak local minimum but the converse is not true.

The following hypotheses will be imposed on Problem 1.1 throughout this chapter, mostly without further mentioning.

Hypothesis 1.6 The functions $g(x,u,t)$ and $f(\varphi,u,t)$ are measurable in t for every fixed $(x,u) \in \mathcal{O}_x \times \mathcal{O}_u$ and $(\varphi,u) \in \mathcal{O}_\varphi \times \mathcal{O}_u$, respectively; they are continuously Fréchet differentiable with respect to (x,u) and (φ,u) respectively, for a.a. $t \in T$. The function $h(x,t)$ is Fréchet differentiable with respect to x and $h(x,t)$, $h_x(x,t)$ are continuous in (x,t).

Hypothesis 1.7 There exists a function $q: R_+ \times R \to R_+$ such that $q(s,\cdot) \in L^1(T;R)$ for all $s \in R_+$, $q(\cdot,t)$ is monotonically increasing for a.a. $t \in T$, and the following conditions hold:

$$|g(x,u,t)| + |g_{x,u}(x,u,t)| \le q(|x| + |u|,t)$$

$$|f(\varphi,u,t)| + |\mathcal{D}_{1,2}f(\varphi,u,t)| \le q(|\varphi| + |u|,t)$$

for all $(x,u) \in \mathcal{O}_x \times \mathcal{O}_u$, $(\varphi,u) \in \mathcal{O}_\varphi \times \mathcal{O}_u$ and a.a. $t \in T$.

Hypothesis 1.8 The sets $\Omega(t) \subset R^m$ are closed and convex.

Hypothesis 1.9 The function p is continuously Fréchet differentiable.

Remark 1.9 It suffices to require the conditions above in an ε-tube around $x^0(t)$, $t \in T$. Note that by continuity $h(x,t)$ and $h_x(x,t)$ are bounded for bounded $|x|$ and $t \in T$.

Remark 1.10 Observe that the assumptions above are stronger than those of Chapter IV in that we require $\Omega(t)$ to be convex and f,g to be differentiable with respect to u. On the other hand, the fundamental extendability Hypothesis IV.1.6 is not needed here.

We reformulate the control Problem 1.1 as an optimization problem in Banach spaces in order to apply the results of sections II.1 and II.2. Among the numerous possibilities for such a reformulation we choose one which appears straightforward and allows us to determine the concrete form of the optimality conditions by using the structural theory of functional differential equations of Chapter III.

There exists an open subset $\tilde{\mathcal{O}}$ of $C(t_1-r,t_2,R^n) \times L^\infty(t_1,t_2;R^m)$ such that $(x^0,u^0) \in \tilde{\mathcal{O}}$ and for $(x,u) \in \tilde{\mathcal{O}}$ one has $x_t \in \mathcal{O}_\varphi$ for all $t \in T$ (this follows from compactness of $\{x_t^0,\ t \in T\}$).

Hence there exists an open set $\mathcal{O} = \mathcal{O}_X \times \mathcal{O}_U \times \mathcal{O}_\varphi \subset C(T;R^n) \times L^\infty(T;R^m) \times C(-r,0;R^n)$ with $(x^o,u^o,\varphi^o) \in \mathcal{O}$ such that for $(x,u,\varphi) \in \mathcal{O}$ the following functional G and the following maps F,H with values in $C(T;R^n)$, and $C(T;R^\ell)$, respectively, are well-defined:

$$G(x,u) := \int_{t_1}^{t_2} g(x(s),u(s),s)ds; \tag{1.2}$$

$$[F(x,u,\varphi)](t) := \varphi(0) + \int_{t_1}^{t} f(x_s,u(s),s)ds, \quad t \in T \tag{1.3}$$

where at the right hand side

$$x(t_1+s) := \varphi(s), \quad s \in [-r,0);$$

$$[H(x)](t) := h(x(t),t), \quad t \in T. \tag{1.4}$$

There is a slight technical difficulty in defining F as above: For $t \in (t_1,t_2+r)$, the function x_t is not necessarily in $C(-r,0;R^n)$, since it may have a jump at $s = t_1-t$; however, one can always extend $f(\cdot,u(t),t)$ smoothly to the space of functions on $[-r,0]$ which are allowed to have a single discontinuity at t_1-t.

We note the following results.

Lemma 1.11 (i) The function F is continuously Fréchet differentiable and $Id - \mathcal{D}_1F(x^o,u^o,\varphi^o)$ is an isomorphism on $C(T;R^n)$.

(ii) The linearized equation

$$x = \mathcal{D}F(x^o,u^o,\varphi^o)(x,u,\varphi) \tag{1.5}$$

is equivalent to

$$x_{t_1} = \varphi, \ \dot{x}(t) = \mathcal{D}_1f(x_t^o,u^o(t),t)x_t + \mathcal{D}_2f(x_t^o,u^o(t),t)u(t), \quad \text{a.a. } t \in T. \tag{1.6}$$

(iii) Equation (1.6) has for every (u,φ) a unique solution x.

(iv) The equation

$$x = F(x,u,\varphi) \tag{1.7}$$

has for given (u,φ) in a neighborhood of (u^o,φ^o) a unique solution $x = x(u,\varphi)$. Without loss of generality let $\mathcal{O}_U \times \mathcal{O}_\varphi$ be this neighborhood.

(v) The solution map

$$S: \mathcal{O}_U \times \mathcal{O}_\varphi \to C(T;R^n)$$

of equation (1.7) is continuously Fréchet differentiable at (u^0,φ^0) and

$$x := DS(u^0,\varphi^0)(u,\varphi)$$

is the solution of (1.6).

<u>Proof:</u> The assertions follow as Proposition III.3.2. □

Define the map

$$P: C(T;R^n)\times C(-r,0;R^n) \to Y$$

by

$$P(x,\varphi) := p(x_{t_2},\varphi), \tag{1.8}$$

where $x(t_1+s) := \varphi(s)$, $s \in [-r,0]$, if $t_2 < t_1+r$.
Let K denote the natural negative cone in $C(T;R^\ell)$.

With these definitions, Problem 1.1 can be reformulated as

<u>Problem 1.12</u> Minimize $G(S(u,\varphi),u)$

over all $(u,\varphi) \in \mathcal{O}_U \times \mathcal{O}_\varphi$

satisfying $P(S(u,\varphi),\varphi) = 0$

$H(S(u,\varphi)) \in K$

$u \in \mathcal{U}_{ad}$.

We note the following local equivalence result.

<u>Lemma 1.13</u> A pair (x^0,u^0) is a weak local solution of Problem 1.1 iff $(u^0,x^0_{t_1})$ is a local solution of Problem 1.12.

2. First Order Necessary Optimality Conditions

We embark to prove first order necessary optimality conditions for Problem 1.12, which yield corresponding results for Problems 1.1 and 1.2.

Let (u^0,φ^0) be a local optimal solution of Problem 1.12 and $x^0 := S(u^0,\varphi^0)$. Problem 1.12 is a special case of Problem II.1.1 in the spaces

$$X := L^\infty(T;R^m)\times C(-r,0;R^n), \quad Z := C(T;R^\ell)$$

and Y as specified in the problem; the constraint set C is

$$C := \mathcal{U}_{ad}\times C(-r,0;R^n) \subset X.$$

First check the required differentiability properties. Note that the derivative of S has already been determined in Lemma 1.11.

Lemma 2.1 (i) The functional G defined in (1.2) is continuously Fréchet differentiable at (x^0,u^0) with partial derivatives

$$\mathcal{D}_1G(x^0,u^0)x = \int_{t_1}^{t_2} g_x(x^0(t),u^0(t),t)x(t)dt$$

$$\mathcal{D}_2G(x^0,u^0)u = \int_{t_1}^{t_2} g_u(x^0(t),u^0(t),t)u(t)dt.$$

(ii) The maps H and P defined in (1.4) and (1.8) are continuously Fréchet differentiable with derivatives

$$[\mathcal{D}H(x^0)x](t) = h_x(x^0(t),t)x(t), \quad t \in T$$

$$\mathcal{D}P(x^0,\varphi^0)(x,\varphi) = \mathcal{D}p(x^0_{t_2},\varphi^0)(x_{t_2},\varphi).$$

Proof: Similar to Proposition III.3.2. □

Next analyse the required constraint qualifications.
Define the "attainability cone" A by

$$A := \{\mathcal{D}p(x^0_{t_2},\varphi^0)(x_{t_2},\varphi) : \varphi \in C(-r,0;R^n) \tag{2.1}$$

$$\text{and there exists } u \in \mathcal{U}_{ad}(u^0) \text{ such that } x_{t_1} = \varphi \text{ and}$$

$$\dot{x}(t) = \mathcal{D}_1f(x^0_t,u^0(t),t)x_t + \mathcal{D}_2f(x^0_t,u^0(t),t)u(t) \text{ a.a. } t \in T\}$$

Lemma 2.2 The following identity holds:

$$A = \{\mathcal{D}P(x^0,\varphi^0)(\mathcal{D}S(u^0,\varphi^0)(u,\varphi),\varphi)) : \varphi \in C(-r,0;R^n),\ u \in \mathcal{U}_{ad}(u^0)\}.$$

Proof: Clear by Lemmas 1.11 and 2.1. □

Note that, by the chain rule, the condition $A = Y$ means that the equality constraint in Problem 1.12 satisfies the regularity condition (II. 1.4).

Lemma 2.3 The inequality constraint in Problem 1.12 satisfies the constraint qualification (II.1.6) if the following condition holds:

There exist $\tilde{u} \in \mathcal{U}_{ad}(u^0)$ and a solution $\tilde{x}$ of (2.2)

$$\dot{\tilde{x}}(t) = \mathcal{D}_1f(x^0_t,u^0(t),t)\tilde{x}_t + \mathcal{D}_2f(x^0_t,u^0(t),t)\tilde{u}(t), \text{ a.a. } t \in T$$

with

$$\mathcal{D}p(x^o_{t_2},\varphi^o)(\tilde{x}_{t_2},\tilde{x}_{t_1}) = 0$$
$$h_x(x^o(t),t)\tilde{x}(t) \in \text{int } R^{\ell}_-(h(x^o(t),t)) \text{ for all } t \in T.$$

Proof: Clear by Lemmas 1.11 and 2.1. □

We are now in a position to apply the abstract first order necessary optimality conditions of Theorem II.1.11 to Problem 1.12. Define the corresponding Lagrangean L as

$$L(u,\varphi,\lambda) := \lambda_o G(S(u,\varphi),u) - y^*P(S(u,\varphi),\varphi) - z^*H(S(u,\varphi)) \qquad (2.3)$$

where $(u,\varphi) \in L^\infty(T;R^m)\times C(-r,0;R^n)$ and $\lambda := (\lambda_o,y^*,z^*) \in R_+\times Y^*\times C(T;R^\ell)^*$.

Theorem 2.4 Let $(u^o,\varphi^o) \in L^\infty(T;R^m)\times C(-r,0;R^n)$ be a local minimum of Problem 1.12, define $x^o := S(u^o,\varphi^o)$, and assume that the attainability cone A defined in (2.1) contains a subspace of finite codimension in Y. Then there exist $0 \neq \lambda = (\lambda_o,y^*,z^*) \in R_+\times Y^*\times C(T;R^\ell)^*$ such that

$$z^*H(x^o) = 0 \text{ and } z^*z \geq 0 \text{ for all negative } z \in C(T;R^\ell) \qquad (2.4)$$
$$\mathcal{D}_1L(u^o,\varphi^o)u \geq 0 \text{ for all } u \in \mathcal{U}_{ad}(u^o) \qquad (2.5)$$
$$\mathcal{D}_2L(u^o,\varphi^o) = 0 \text{ in } C(-r,0;R^n)^*.$$

If the attainability cone A satisfies $A = Y$ and (2.2) holds, then $\lambda_o \neq 0$.

Proof: Lemmas 1.11 and 2.1 - 2.3 show that Theorem II.1.11 is applicable. Hence there exists $0 \neq \lambda = (\lambda_o,y^*,z^*) \in R_+\times Y^*\times Z^*$ satisfying (2.4) and, by the chain rule, also

$$\mathcal{D}_{1,2}L(u^o,\varphi^o,\lambda)(u,\varphi)$$
$$= \lambda_o\mathcal{D}G(x^o,u^o)(\mathcal{D}S(u^o,\varphi^o)(u,\varphi),u) - y^*\mathcal{D}P(x^o,\varphi^o)(\mathcal{D}S(u^o,\varphi^o)(u,\varphi),\varphi) \geq 0$$

for all $u \in \mathcal{U}_{ad}(u^o)$ and $\varphi \in C(-r,0;R^n)$.

Looking at φ and u separately, one deduces (2.5) and (2.6). □

Note that (2.5) and (2.6) are equivalent to

$$[\lambda_o\mathcal{D}_1G(x^o,u^o) - y^*\mathcal{D}_1P(x^o,\varphi^o) - z^*\mathcal{D}H(x^o)]\mathcal{D}_1S(u^o,\varphi^o)u \qquad (2.7)$$
$$+ \lambda_o\mathcal{D}_2G(x^o,u^o)u \geq 0$$

$$[\lambda_o\mathcal{D}_1G(x^o,u^o) - y^*\mathcal{D}_1P(x^o,\varphi^o) - z^*\mathcal{D}H(x^o)]\mathcal{D}_2S(u^o,\varphi^o)\varphi \qquad (2.8)$$
$$- y^*\mathcal{D}_2P(x^o,\varphi^o)\varphi = 0.$$

As in Chapter IV it is convenient to rewrite the optimality conditions (2.5) and (2.6) with the help of an adjoint equation. Again, let $\Phi(t,s)$, $t \geq s$, denote the family of evolution operators associated with the linear retarded equation

$$\dot{x}(t) = \mathcal{D}_1 f(x_t^o, u^o(t), t)x_t, \qquad t \in R. \tag{2.9}$$

Lemma 1.11 implies that $\mathcal{D}_1 S(u^o,\varphi^o)u$ and $\mathcal{D}_2 S(u^o,\varphi^o)\varphi$ are the solution of

$$x_{t_1} = 0, \; \dot{x}(t) = \mathcal{D}_1 f(x_t^o, u^o(t), t)x_t + \mathcal{D}_2 f(x_t^o, u^o(t), t)u(t) \quad \text{a.a. } t \in T \tag{2.10}$$

$$x_{t_1} = \varphi, \; \dot{x}(t) = \mathcal{D}_1 f(x_t^o, u^o(t), t)x_t \quad \text{a.a. } t \in T, \tag{2.11}$$

respectively.

<u>Remark 2.5</u> The weak variations of the control u considered here lead to an inhomogenity in the variational equation (2.10). This is in contrast with strong variations which lead to an initial value of the variational equation, Lemma IV.2.3.

By the variation of constants formula one obtains for $t \in T$

$$[\mathcal{D}_1 S(u^o,\varphi^o)u]_t = \int_{t_1}^{t} \Phi(t,s)X_o \mathcal{D}_2 f(x_s^o, u^o(s), s)u(s)ds \tag{2.12}$$

$$[\mathcal{D}_2 S(u^o,\varphi^o)\varphi]_t = \Phi(t,t_1)\varphi.$$

First we will discuss the periodic Problem 1.2, where

$$y^* \in Y^* = C(-r,0;R^n)^* = NBV(-r,0;R^n)$$

$$\mathcal{D}P(x^o,\varphi^o)(x,\varphi) = x_{t_2} - \varphi$$

furthermore $z^* \in C(T,R^\ell)^*$ will be identified with $\mu = (\mu_i)_{i=1\ldots\ell}$, where μ_i is a regular Borel measure on T.

Note that

$$\begin{aligned} z^*\mathcal{D}H(x^o)\mathcal{D}_2 S(u^o,\varphi^o)\varphi &= z^* h_x(x^o(\cdot),\cdot)[\Phi(\cdot,t_1)\varphi](0) \\ &= \int_{t_1}^{t_2} d\mu(t)^T h_x(x^o(t),t)[\Phi(t,t_1)\varphi](0). \end{aligned} \tag{2.13}$$

By Lemma 2.1 equation (2.8) is equivalent to

$$\begin{aligned} y^*\varphi = {} & \lambda_0/(t_2-t_1) \int_{t_1}^{t_2} g_x(x^o(t),u^o(t),t)[\Phi(t,t_1)\varphi](0)dt \\ & - y^*\Phi(t_2,t_1)\varphi - \int_T [d\mu^T(t)]h_x(x^o(t),t)[\Phi(t,t_1)\varphi](0) \end{aligned} \tag{2.14}$$

for all $\varphi \in C(-r,0;R^n)$.

If $t_2 > t_1+r$ this implies as Lemma IV.2.7 the existence of $\psi \in NBV(0,r;R^n)$ with

$$y^* = F^*(t_2)\psi, \tag{2.15}$$

where $F(t)$ is the structural operator associated with (2.9).
The "adjoint equation" in $NBV(0,r;R^n)$ is

$$y^s = \Psi^*(t_2,s)\psi - \int_s^{t_2} \Psi^*(t,s)Y_o\lambda_0/(t_2-t_1)g_x(x^0(t),u^o(t),t)dt \tag{2.16}$$
$$- \int_{[s,t_2]} [d\mu^T(t)]\Psi^*(t,s)Y_o h_x(x^0(t),t), \quad t \in T.$$

Equivalently, one has in R^n (cp. Proposition III.1.8)

$$y(s) - y(t_2) = - \int_t^{t_2+r} [\eta^T(\alpha,s-\alpha)-\eta^T(\alpha,t_2-\alpha)]y(\alpha)d\alpha \tag{2.17}$$
$$+ \lambda_0/(t_2-t_1) \int_s^{t_2} g_x(x^0(t),u^o(t),t)dt$$
$$+ \int_{[s,t_2]} [d\mu^T(t)]h_x(x^0(t),t), \quad s \in T,$$
$$y^{t_2} = \psi.$$

Arguing as for (IV.2.18) one gets the following result.

Lemma 2.6 (i) Let y be a solution of (2.17) with $y^{t_1} = y^{t_2}$. Then $y^* := F^*(t_2)y^{t_2}$ satisfies (2.14).

(ii) Conversely, suppose that y^* in the range of $F^*(t_2)$ satisfies (2.14). Then there exists $\psi \in NBV(0,r;R^n)$ with $y^* = F^*(t_2)\psi$ such that the corresponding solution y of (2.17) satisfies $y^{t_1} = y^{t_2}$.

Next we discuss the required constraint qualification.

Lemma 2.7 In the periodic Problem 1.2, the attainability cone is given by

$$A = \{x_{t_2}-\varphi : \varphi \in C(-r,0;R^n) \text{ and there exist } \alpha > 0 \tag{2.18}$$
$$\text{and } u \in L^\infty(T;R^m) \text{ with } \alpha u^o(t) + u(t) \in \alpha\Omega(t) \text{ a.e.}$$
$$\text{such that } x \text{ is the corresponding solution of (1.5)}\}.$$

The cone A contains a subspace of finite codimension in $Y = C(-r,0;R^n)$.

Proof: One only has to prove that A is given by (2.18) contains a subspace of finite codimension in $C(-r,0;R^n)$.

There exists $m \geq 1$ such that $(t_2-t_1)m \geq r$, and hence, by periodicity

$$\Phi(t_2+(t_2-t_1)m,t_1) = \Phi(t_2,t_1)^{m+1}.$$

Thus $\Phi(t_2,t_1)^m$ is compact and the range of

$$[Id-\Phi(t_2,t_1)^m] = [Id-\Phi(t_2,t_1)][Id+\Phi(t_2,t_1)+\ldots+\Phi(t_2,t_1)^{m-1}]$$

has finite codimension. Hence this is also true for $Id - \Phi(t_2,t_1)$ and the assertion follows. □

Proposition 2.8 Let $t_2 \geq t_1+r$. Then each of the following conditions implies that $A = C(-r,0;R^n)$, where A is given by (2.18):

The equation (2.19)

$$\dot{x}(t) = \mathcal{D}_1 f(x_t^o,u^o(t),t)x_t, \quad t \geq t_1$$

has only the trivial (t_2-t_1)-periodic solution.

The linearized system (2.20)

$$\dot{x}(t) = \mathcal{D}_1 f(x_t^o,u^o(t),t)x_t + \mathcal{D}_2 f(x_t^o,u^o(t),t)u(t), \quad \text{a.a. } t \in [t_1,t_2]$$

$$x_{t_1} = 0$$

where $u \in L^\infty(T;R^n)$ satisfies the positivity constraint

$$u(t) = \alpha(v(t)-u^o(t)), \quad \text{a.a. } t \in T,$$

with $\alpha \geq 0$, $v(t) \in \Omega(t)$ a.a. $t \in T$,

is approximately controllable to $C(-r,0;R^n)$, i.e. the set of all x_{t_2}, x a trajectory of the system above, contains a *dense subspace* of $C(-r,0;R^n)$.

Proof: The proof of the first assertion follows at once from the arguments in the proof of Lemma 2.7.

Concerning sufficiency of (2.20) observe that A is the sum of range $[Id-\Phi(t_2-t_1)]$ and the attainability subspace of the system in (2.20). Thus the assertion follows, since range $[Id-\Phi(t_2,t_1)]$ has finite codimension and the sum of such a subspace and a dense subspace concides with the whole space. □

We note that for the periodic Problem 1.2, condition (2.2) has the following form:

$$\begin{aligned}&\text{There exist } \tilde{u} \in \mathcal{U}_{ad}(u^o) \text{ and a solution } \tilde{x} \text{ of} && (2.21)\\ &\dot{\tilde{x}}(t) = \mathcal{D}_1 f(x_t^o, u^o(t), t)\tilde{x}_t + \mathcal{D}_2 f(x_t^o, u^o(t), t)\tilde{u}(t), \text{ a.a. } t \in T\\ &\text{with } \tilde{x}_{t_1} = \tilde{x}_{t_2} \text{ and}\\ &h_x(x^o(t),t)\tilde{x}(t) \in \operatorname{int} R_-^{\ell}(h(x^o(t),t)) \text{ for all } t \in T.\end{aligned}$$

With these preparations, the proof of the following *Weak or Local Maximum Principle* for the periodic Problem 1.2 will be quite easy.

<u>Theorem 2.9</u> Let $t_2 \geq t_1 + r$ and suppose that $(x^o, u^o) \in C(-r,0;R^n) \times L^\infty(t_1,t_2;R^m)$ is a weak local minimum of Problem 1.2 satisfying Hypotheses 1.6 - 1.8. Then there exist $\lambda_o \geq 0$, non-negative regular Borel measures μ_i on T supported on the sets $\{t : h^i(x^o(t),t) = 0\}$, $i = 1,\dots,\ell$, and a (t_2-t_1)-periodic solution y of the adjoint equation $(\mu := (\mu_i))$

$$y(s) - y(t_2) + \int_s^{t_2+r} [\eta^T(\alpha, s-\alpha) - \eta^T(\alpha, t_2-\alpha)]y(\alpha)d\alpha \tag{2.22}$$
$$= \lambda_o \int_s^{t_1} g_x(x^o(t),u^o(t),t)dt + \int_{[s,t_2]} [d\mu^T(t)]h_x(x^o(t),t), \quad s \in T,$$

such that $(\lambda_o, y^{t_2}, \mu) = \lambda \neq 0$ in $R_+ \times NBV(0,r;R^n) \times C(T;R^{\ell})^*$ and

$$[\lambda_o g_u(x^o(s),u^o(s),s) + y(s)^T f_u(x^o(s),u^o(s),s)][\omega - u^o(s)] \geq 0 \tag{2.23}$$
$$\text{for all } \omega \in \Omega(s) \text{ and a.a. } s \in T;$$

furthermore, if the attainability cone A specified in (2.18) satisfies $A = C(-r,0;R^n)$ and condition (2.21) holds, then $\lambda_o \neq 0$; here η is given by the representation

$$\mathcal{D}_1 f(x_s^o, u^o(s), s)\varphi = \int_{-r}^{0} [d_s\eta(s,t)]\varphi(t), \quad \varphi \in C(-r,0;R^n). \tag{2.24}$$

<u>Proof:</u> This result follows from Theorem 2.4, using Lemmas 2.6 and 2.7, and arguments similar to those of section IV.2. We only note that the pointwise form (2.23) of the local maximum condition follows from the integrated form

$$\int_{t_1}^{t_2} [\lambda_o g_u(x^o(s),u^o(s),s) + y(s)^T f_u(x^o(s),u^o(s),s)]u(s)ds \geq 0 \tag{2.25}$$

for all $u \in \mathcal{U}_{ad}(u^o)$

as Warga [1972, Theorem VI.2.3]. □

Remark 2.10 Observe that the local minimum condition (2.23) is a first order necessary condition for a local minimum of

$$\lambda_0 g(x^0(s),\omega,s) + y(s)^T f(x_s^o,\omega,s)$$

at $\omega = u^o(s)$ in $\Omega(s)$. One should not expect more for first order weak variations as considered above.

Remark 2.11 Suppose that no state and control constraints are present and that $A = C(-r,0;R^n)$. Then, given $\lambda_0 > 0$, the function y is uniquely determined by conditions (2.22) and (2.23). However, the periodic solution of (2.22) may *not* be unique, since the condition $A = C(-r,0;R^n)$ may be satisfied due to (2.20) while (2.19) is violated.

Remark 2.12 An alternative treatment of the constraint (2.24) would involve the non-differentiable constraint

$$\max_{t\in T} h(x(t)) \le 0$$

(cf. Ioffe/Tikhomirov [1979]), using e.g. methods in Ben-Tal/Zowe [1982].

Remark 2.13 For functional differential systems with state constraints, but without periodicity condition, Buehler [1976] derived a maximum principle, using Neustadt's abstract variational methods; see also Neustadt [1976]. For problems with fixed boundary condition Kim/Bien [1981] consider state constraints in a maximum principle using methods due to Makowski/Neustadt [1974].

We turn to a brief discussion of fixed boundary value problems, where p has the form

$$p(x_{t_1},x_{t_2}) = (x_{t_2} - \varphi^2, x_{t_1} - \varphi^1) \tag{2.26}$$

with $\varphi^1,\varphi^2 \in C(-r,0;R^n)$ fixed; no state constraint will be imposed. This discussion serves two purposes: At one hand, an interesting theory is associated with this control problem; and at the other hand, the results obtainable for this problem are of very limited scope. This is sharp contrast to the optimal periodic problem and it is hoped that exposure to the atrocities associated with (2.26) will lead the reader to appreciate the nice theory associated with the periodic problem.

Gabasov/Kirillova [1981] mentioned the problem with fixed (function space) end points as one of the open relevant problems in optimal control of delay equations.

Suppose p is given by (2.26) and look at the Banach space reformulation, Problem 1.12.

Since $x_{t_1} = \varphi^1$ is fixed, it suffices to minimize over all $u \in \mathcal{U}_{ad}$: Let $S(u) := S(u,\varphi^0)$.

For an application of Theorem 2.4, we have yet to specify Y and to analyze its subset

$$A = \{(\mathcal{D}S(u^0)u)_{t_2} : u \in \mathcal{U}_{ad}(u^0)\}.$$

In order to get the optimality conditions with $\lambda_0 = 1$, we need, at least, that

$$Y = \{(\mathcal{D}S(u^0)u)_{t_2} : u \in L^\infty(T;R^m)\}. \tag{2.27}$$

But $x := \mathcal{D}S(u^0)u$ is the solution of

$$x_{t_1} = 0,\ \dot{x}(t) = \mathcal{D}_1 f(x_t^0,u^0(t),t)u(t) + \mathcal{D}_2 f(x_t^0,u^0(t),t)u(t), \quad \text{a.a. } t \in T. \tag{2.28}$$

Hence (2.27) is equivalent to

$$Y = A_0^\infty := \{x_{t_2} : \text{there exists } u \in L^\infty(T;R^m) \text{ such that x solves (2.28)}\}$$

A_0^∞ is the subspace attainable from zero with L^∞-controls. But for $t_2 > t_1+r$, $A_0^\infty \subset W_\infty^{(1)}(-r,0;R^n)$, which precludes the choices $Y = C(-r,0;R^n)$ or $Y = W_p^{(1)}(-r,0;R^n)$, $1 \le p < \infty$.

The choice $Y = W_\infty^{(1)}(-r,0;R^n)$ is possible. However, a result due to Jacobs/Kao [1972], see also Banks/Jacobs/Langenhop [1974,1975], says that for $1 \le p \le \infty$ the equality

$$A_0^p := \{x_{t_2} : \text{there exists } u \in L^p(T;R^m) \text{ such that x solves (2.28)}\} \tag{2.29}$$
$$= W_p^{(1)}(-r,0;R^n)$$

implies

$$\operatorname{rank} \mathcal{D}_2 f(x_t^0,u^0(t),t) = n \quad \text{for a.a. } t \in [t_2-r,t_2]. \tag{2.30}$$

This is a very strong condition requiring in particular that the number of control variables is not less than the number n of the state variables (cp. also Banks [1972], Banks/Manitius [1974], Banks/Kent [1972],

Olbrot [1977] and Bien [1975], Bien/Chyung [1980]; the latter use the approach due to Makowski/Neustadt [1974] for mixed phase/control equality constraints. By differentiation, the end condition can be transformed into a constraint of this type). See Section VI.3 for a further discussion of this problem using relaxed controls.

Remark 2.14 Jacobs/Kao [1972], Banks/Jacobs [1973], Colonius/Hinrichsen [1978] develop the theory for L_2-controls and $Y = W_2^{(1)}(-r,0;R^n)$. Here pointwise control constraints cannot be allowed and the control must appear affinely linear in the equation; if the latter condition does not hold, a result by Vainberg [1952] implies that Fréchet differentiability is *never* satisfied (contrary to assertions in Jacobs/Kao [1972], Das [1975]). For *linear*, autonomous delay systems

$$\dot{x}(t) = A_0x(t) + A_1x(t-r) + B_0u(t), \tag{2.31}$$

A_0, A_1, B_0 matrices of appriate dimensions, one has to require that A_0^2 is closed in $W_2^{(1)}(-r,0;R^n)$, Banks/Jacobs [1973]. This is *equivalent*, Kurcyusz/Olbrot [1977], to the condition

$$\operatorname{Im} A_iA_0^iB_0 \subset \operatorname{Im} B_0, \quad i = 0,1,\dots,n-1. \tag{2.32}$$

This condition is satisfied for n-th order scalar delay equations, Jacobs [1972], but for general linear equations of the form (2.31) it is (in any reasonable sense) a non-generic condition; it is not preserved under small perturbations of the entries in A_0, A_1, B_0.

For related work see Banks/Jacobs/Langenhop [1975], Bartosiewicz [1979, 1984], Bartosiewicz/Sienkiewicz [1984], Jakubczyk [1978].

Remark 2.15 In order to avoid condition (2.32), a different formalization of "fixed boundary value problems" has been proposed in Colonius [1982c,1984]. Here the end condition fixes $F(t_2)x_{t_2}$, instead of x_{t_2}. This formulation for linear problems yields satisfactory results in the case of unbounded L_2-controls. In particular for systems of the form (2.31), the required closedness condition (for a "small attainability subspace") is implied by

$$\operatorname{Im} A_1A_0^iB_0 \subset \operatorname{Im} A_1B_0, \quad i = I,1,\dots,n-1. \tag{2.33}$$

This condition implies rank $A_1 \le m$, if the pair (A_0,B_0) is controllable. However, it is much less restrictive than (2.32): it holds, in particular, "generically" (in an algebraic sense) in the variety

$$V = \{(A_0,A_1,B_0) \in R^{n\times n}\times R^{n\times n}\times R^{n\times m}: \text{rank } A_1 \le m\}.$$

Thus, in this formulation, a maximum principle holds generically if the following "law of requisite variety in control" is satisfied: The number of linearly independent delay terms, that is rank A_1, must not exceed the number m of control inputs.

Remark 2.16 Olbrot [1976] replaces the fixed end condition by an approximate end condition of the form $|x_{t_2}-\varphi^2| \le \varepsilon, \quad \varepsilon > 0$.

Remark 2.17 For neutral functional differential equations, the required conditions on the system matrices are much less restrictive (cp. in this context Jacobs/Langenhop [1978], Utthoff [1979], Salamon [1984]).

3. Second Order Necessary Optimality Conditions

This section presents necessary optimality conditions of second order, which are stronger than the Legendre-Clebsch Condition in Corollary IV.2.11.

For an application of the abstract result, Corollary II.2.12, we have to complement the standing Hypotheses 1.6 - 1.9 by the following second order differentiability conditions which are required throughout this section, mostly without further mentioning.

Hypothesis 3.1 The functions $g(x,u,t)$, $f(\varphi,u,t)$, $h(x,t) = (h^i(x,t))$, and $p(x,\varphi)$ are twice continuously Fréchet differentiable with respect to (x,u), (φ,u), x, and (x,φ), respectively.

Hypothesis 3.2 For all $(x,u) \in \mathcal{O}_x \times \mathcal{O}_u$, $(\varphi,u) \in \mathcal{O}_\varphi \times \mathcal{O}_u$ and a.a. $t \in T$

$$|g_{xx}(x,u,t)| + |g_{xu}(x,u,t)| + |g_{uu}(x,u,t)| \le q(|x|+|u|,t)$$
$$|\mathcal{D}_1\mathcal{D}_1 f(\varphi,u,t)| + |\mathcal{D}_1\mathcal{D}_2 f(\varphi,u,t)| + |\mathcal{D}_2\mathcal{D}_2 f(\varphi,u,t)| \le q(|\varphi|+|u|,t),$$

where q is as in Hypothesis 1.7; for $i = 1,\dots,\ell$, the functions $h^i_{xx}(x,t)$ are continuous with respect to (x,t).

Note that $|\mathcal{D}_1\mathcal{D}_1 f(\varphi,u,t)|$ is the norm of the second derivative of $f(\varphi,u,t)$ with respect to φ, i.e. the norm of $\mathcal{D}_1\mathcal{D}_1 f(\varphi,u,t) \in L(C(-r,0;R^n),L(C(-r,0;R^n),R^n))$ (cf. Berger [1977, section 2.1.E]).

These hypotheses imply the required differentiability properties, as stated in the following lemmas.

Lemma 3.3 The functions G and H defined in (1.2) and (1.4) are twice continuously Fréchet differentiable with

$$\mathcal{D}\mathcal{D}G(x^0,u^o)((x,u) = \int_{t_1}^{t_2}[x(t)^T g_{xx}(x^0(t),u^o(t),t)x(t)$$

$$+ 2x(t)^T g_{xu}(x^0(t),u^o(t),t)u(t) + u(t)^T g_{uu}(x^0(t),u^o(t),t)u(t)]dt.$$

$$[\mathcal{D}\mathcal{D}H(x^0)(x,x)](t) = \sum_{i=1}^{\ell} x(t)^T h^i_{xx}(x^0(t),t)x(t), \quad t \in T.$$

Proof: This is a simple consequence of the hypotheses. □

Lemma 3.4 The function F defined in (1.3) is twice continuously Fréchet differentiable with

$$[\mathcal{D}\mathcal{D}F(x^0,u^o,\varphi^0)((x,u,\varphi),(x,u,\varphi))](t) = \int_{t_1}^{t} [\mathcal{D}_1\mathcal{D}_1 f(x_s^o,u^o(s),s)(x_s^\varphi,x_s^\varphi)$$

$$+ 2\mathcal{D}_1\mathcal{D}_2 f(x_s^o,u^o(s),s)(x_s^\varphi,u(s)) + \mathcal{D}_2\mathcal{D}_2 f(x_s^o,u^o(s),s)(u(s),u(s))]ds,$$
$$t \in T,$$

where $x_{t_1}^\varphi := \varphi$ and $x^\varphi(t) := x(t)$, for $t \in T$.

Proof: Observe that e.g.

$$[\mathcal{D}_3\mathcal{D}_3 F(x^0,u^o,\varphi^0)(\varphi,\varphi)] = \int_{t_1}^{t} [\mathcal{D}_1\mathcal{D}_1 f(x_s^o,u^o(s),s)(x_s,x_s)ds, \quad t \in T,$$

where $x_{t_1} := \varphi$ and $x(t) = 0$ for $t \in T$.
Similar formulae for the other partial derivatives, continuity, and linearity give the assertion. □

The following lemma is crucial.

Lemma 3.5 The solution operator S of equation (1.4) is twice continuously Fréchet differentiable and

$$\xi := \mathcal{D}\mathcal{D}S(u^o,\varphi^0)((u,\varphi),(u,\varphi))$$

is the unique solution of

$$\xi = \mathcal{D}_1 F(x^0,u^o,\varphi^0)\xi + \mathcal{D}\mathcal{D}F(x^0,u^o,\varphi^0)((x,u,\varphi),(x,u,\varphi)),$$

where $x^0 := S(u^o,\varphi^0)$ and $x := \mathcal{D}S(u^o,\varphi^0)(u,\varphi)$.

Proof: The implicit function theorem implies that S is twice continuously Fréchet differentiable, since F is. We compute the second derivative by applying the second order chain rule, Proposition II.2.5, to

$$S(u,\varphi) - F(S(u,\varphi),u,\varphi) = 0$$

and obtain

$$\begin{aligned}
&\mathcal{D}\mathcal{D}S(u^0,\varphi^0)((u,\varphi),(u,\varphi)) \\
&- \mathcal{D}\mathcal{D}F(S(u^0,\varphi^0),u^0,\varphi^0)[(\mathcal{D}S(u^0,\varphi^0)(u,\varphi),u,\varphi),(\mathcal{D}S(u^0,\varphi^0)(u,\varphi),u,\varphi)] \\
&- \mathcal{D}F(S(u^0,\varphi^0),u^0,\varphi^0)[\mathcal{D}\mathcal{D}S(u^0,\varphi^0)((u,\varphi),(u,\varphi)),0,0] \\
&= \mathcal{D}\mathcal{D}S(u^0,\varphi^0)((u,\varphi),(u,\varphi)) - \mathcal{D}\mathcal{D}F(x^0,u^0,\varphi^0)[(x,u,\varphi),(x,u,\varphi)] \\
&- \mathcal{D}_1F(x^0,u^0,\varphi^0)\mathcal{D}\mathcal{D}S(u^0,\varphi^0)((u,\varphi),(u,\varphi)).
\end{aligned}$$

□

The proof of the next lemma proceeds similarly and will be omitted.

Lemma 3.6 The maps

$(u,\varphi) \to G(S(u,\varphi),u), (u,\varphi) \to H(S(u,\varphi))$, and $(u,\varphi) \to P(S(u,\varphi),\varphi)$

are twice continuously Fréchet differentiable with derivatives given by

$$\begin{aligned}
&\mathcal{D}\mathcal{D}G(S(u^0,\varphi^0),u^0)((u,\varphi),(u,\varphi)) \\
&= \mathcal{D}\mathcal{D}G(x^0,u^0)((x,u),(x,u)) + \mathcal{D}_1G(x^0,u^0)\xi;
\end{aligned} \tag{3.1}$$

(here $\mathcal{D}\mathcal{D}G(x^0,u^0)$ denotes the second derivative of G with respect to (x,u))

$$\mathcal{D}\mathcal{D}H(S(u^0,\varphi^0))((u,\varphi),(u,\varphi)) = \mathcal{D}\mathcal{D}H(x^0)(x,x) + \mathcal{D}H(x^0)\xi; \tag{3.2}$$

$$\begin{aligned}
&\mathcal{D}\mathcal{D}P(S(u^0,\varphi^0),\varphi^0)((u,\varphi),(u,\varphi)) \\
&= \mathcal{D}\mathcal{D}P(x^0,\varphi^0)((x,\varphi),(x,\varphi)) + \mathcal{D}_1P(x^0,\varphi)\xi;
\end{aligned} \tag{3.3}$$

where $x^0 := S(u^0,\varphi^0)$, $x := \mathcal{D}S(u^0,\varphi^0)(u,\varphi)$

and $\xi := \mathcal{D}\mathcal{D}S(u^0,\varphi^0)((u,\varphi),(u,\varphi))$.

We obtain the following second order necessary conditions, which are analogous to the first order conditions of Theorem 2.4.

Theorem 3.7 Suppose that $(u^0,\varphi^0) \in L^\infty(T;R^m) \times C(-r,0;R^n)$ is a local minimum for Problem 1.12, define $x^0 = S(u^0,\varphi^0)$ and assume that the attainability cone A defined in (2.1) contains a subspace of finite codimension in Y.

Then for every pair (u,φ) with

$$\mathcal{D}G(x^0,u^o)(x,u) \le 0, \quad \mathcal{D}H(x^0)x \in K(H(x^0)) \tag{3.4}$$

$$\mathcal{D}_1P(x^0,\varphi^0)x + \mathcal{D}_2P(x^o,\varphi^o)\varphi = 0, \quad u \in \mathcal{U}_{ad}(u^o)$$

where $x := \mathcal{D}S(u^o,\varphi^0)(u,\varphi)$,
there exist $0 \neq \lambda = (\lambda_0, y^*, z^*) \in R_+ \times Y^* \times C(T;R^\ell)^*$ with

$$z^*z \ge 0 \quad \text{for all negative} \quad z \in C(T;R^\ell), \tag{3.5}$$

$$z^*H(x^0) = z^*\mathcal{D}H(x^0)x = 0$$

such that for L defined in (2.3)

$$\mathcal{D}_1L(u^o,\varphi^0,\lambda)u' \ge 0 \quad \text{for all} \quad u' \in \mathcal{U}_{ad}(u^o), \tag{3.6}$$

$$\mathcal{D}_2L(u^o,\varphi^0,\lambda) = 0 \quad \text{in} \quad C(-r,0;R^n)^*;$$

$$\mathcal{D}_{1,2}\mathcal{D}_{1,2}L(u^o,\varphi^0,\lambda)((u,\varphi),(u,\varphi)) \ge 0. \tag{3.7}$$

If the attainability cone A satisfies $A = Y$, and condition (2.2) holds, then $\lambda_0 \neq 0$.

<u>Proof:</u> By Lemmas 3.3 - 3.6 the assumptions of Corollary II.2.12 are satisfied. This yields the assertions. □

Note first that here, by Lemma 2.7, A always contains a subspace of finite codimension in $Y = C(-r,0;R^n)$. Furthermore, by (3.3),

$$\mathcal{D}\mathcal{D}P(S(u^o,\varphi^0),\varphi^0)((u,\varphi),(u,\varphi)) = \mathcal{D}_1P(x^0,\varphi^0)\xi = \xi_{t_2}$$

since P is linear.
Thus by Lemma 3.3 - 3.6

$$\begin{aligned}\mathcal{D}_{1,2}\mathcal{D}_{1,2}&L(u^o,\varphi^0,\lambda)((u,\varphi),(u,\varphi)) \\ &= \lambda_o\mathcal{D}\mathcal{D}G(x^0,u^o)((x,u),(x,u)) - z^*\mathcal{D}\mathcal{D}H(x^0)(x,x) \\ &+ \lambda_0\mathcal{D}_1G(x^0,u^o)\xi - y^*\xi_{t_2} - z^*\mathcal{D}H(x^o)\xi.\end{aligned} \tag{3.8}$$

One can write

$$[\mathcal{D}\mathcal{D}F(x^0,u^o,\varphi^0)((x,u,\varphi),(x,u,\varphi))](t) = \int_{t_1}^{t} f_2(\sigma)d\sigma, \quad t \in T;$$

where f_2 is given by the integrand in Lemma 3.4. Hence Lemma 3.5 and the variation of constants formula give

$$\xi_t = \int_{t_1}^{t} \Phi(t,\sigma)X_o f_2(\sigma)d\sigma, \quad t \in T.$$

Lemma 2.1 and (2.13) imply

$$\lambda_0 \mathcal{D}_1 G(x^o,u^o)\xi - y^*\xi_{t_2} - z^*\mathcal{D}H(x^o)\xi$$

$$= \lambda_0 \int_{t_1}^{t_2} g_x(x^o(t),u^o(t),t)[\int_{t_1}^{t} \Phi(t,\sigma)X_o f_2(\sigma)d\sigma](0)dt$$

$$- y^* \int_{t_1}^{t_2} \Phi(t_2,\sigma)X_o f_2(\sigma)d\sigma$$

$$- \int_{[t_1,t_2]} [d\mu^T(t)]h_x(x^o(t),t)[\int_{t_1}^{t} \Phi(t,\sigma)X_o f_2(\sigma)d\sigma)](0).$$

Now, using the properties of Φ as in the proof of Theorem 2.9, one shows that this equals

$$\int_{t_1}^{t_2} y^s X_o f_2(s)ds,$$

where y^s, $s \in T$ is the solution of the adjoint equation (2.16). Plugging this into (3.8) one gets

$$\mathcal{D}_{1,2}\mathcal{D}_{1,2}L(u^o,\varphi^o,\lambda)((u,\varphi),(u,\varphi)) \tag{3.9}$$

$$= \lambda_0 \mathcal{D}\mathcal{D}G(x^o,u^o)((x,u),(x,u)) - z^*\mathcal{D}\mathcal{D}H(x^o)(x,x)$$

$$+ \int_{t_1}^{t_2} y(s)^T f_2(s)ds.$$

In order to write down the concrete form of the optimality conditions, it is convenient to use the following notation (cp. Remark IV.2.9). Write for $\varphi \in C(-r,0;R^n)$, $\omega \in R^m$,

$$\lambda(t) = (\lambda_0, y(t),\mu) \in R\times R^n\times C(T;R^\ell)^*, \quad t \in T$$

$$\begin{aligned} H(\varphi,\omega,\lambda(t),t) := \lambda_0 g(\varphi(0),\omega,t) &+ y^T(t)f(\varphi,\omega,t) \\ &+ [d\mu^T(t)]h(\varphi(t),t) \end{aligned} \tag{3.10}$$

and use $\mathcal{D}_1\mathcal{D}_1 H(x_t^o,u^o(t),\lambda(t),t)$ as a shorthand for

$$\begin{aligned} \lambda_o x(t)^T g_{xx}(x^o(t),u^o(t),t)x(t) &+ y(t)^T \mathcal{D}_1\mathcal{D}_1 f(x_t^o,u^o(t),t)(x_t,x_t) \\ &+ [d\mu^T(t)]x(t)^T h_{xx}(x^o(t),t)x(t); \end{aligned}$$

similarly for $\mathcal{D}_1\mathcal{D}_2 H$, $\mathcal{D}_2 H$, and $\mathcal{D}_2\mathcal{D}_2 H$.

We arrive at the following second order necessary optimality conditions for the periodic Problem 1.2.

<u>Theorem 3.8</u> Suppose $t_2 \geq t_1+r$ and let $(x^0,u^o) \in C(t_1-r,t_2;R^n)\times L^\infty(t_1,t_2;R^m)$ be a weak local minimum of Problem 1.2 such that Hypotheses 1.6 - 1.9 and 3.1, 3.2 are satisfied. Let $(x,u) \in C(t_1-r,t_2;R^n)\times L^\infty(t_1,t_2;R^m)$ be given with

$$\int_{t_1}^{t_2} [g_x(x^0(t),u^o(t),t)x(t) + g_u(x^0(t),u^o(t),t)u(t)]dt \leq 0 \tag{3.11}$$

$$x_{t_1} = x_{t_2}, \; \dot{x}(t) = \mathcal{D}_1 f(x_t^o,u^o(t),t)x_t + \mathcal{D}_2 f(x_t^o,u^o(t),t)u(t), \quad \text{a.a. } t \in T \tag{3.12}$$

$$h(x^0(t),t) + h_x(x^0(t),t)x(t) \in R_-^\ell \text{ all } t \in T \tag{3.13}$$

$$u \in \mathcal{U}_{ad}(u^o) \tag{3.14}$$

Then there exist $\lambda_0 \geq 0$, non-negative regular Borel measures μ_i on T supported on the sets

$\{t:h^i(x^0(t),t) = 0\} \cap \{t:h_x^i(x^0(t),t)x(t) = 0\}$, $i = 1,\ldots,\ell$,

and a (t_2-t_1)-periodic solution y of the adjoint equation (2.22) such that $0 \neq \lambda = (\lambda_0, y^{t_2}, \mu)$ in $R\times NBV(0,r;R^n)\times C(T;R^\ell)^*$ and for $\lambda(t) = (\lambda_0, y(t), \mu)$

$$\mathcal{D}_2 H(x_t^o,u^o(t),\lambda(t),t)[\omega-u^o(t)] \geq 0 \quad \text{for all } \omega \in \Omega(t) \text{ and a.a. } t \in T \tag{3.15}$$

$$\begin{aligned}\int_{t_1}^{t_2} \{&\mathcal{D}_1\mathcal{D}_1 H(x_t^o,u^o(t),\lambda(t),t)(x_t,x_t) \\ &+ 2\mathcal{D}_1\mathcal{D}_2 H(x_t^o,u^o(t),\lambda(t),t)(x_t,u(t)) \\ &+ \mathcal{D}_2\mathcal{D}_2 H(x_t^o,u^o(t),\lambda(t),t)(u(t),u(t))\}dt \geq 0.\end{aligned} \tag{3.16}$$

If the attainability cone A specified in (2.18) satisfies $A = C(-r,0;R^n)$ and condition (2.21) holds, then $\lambda_0 \neq 0$.

Proof: This follows from Theorem 3.7 and the analysis above. □

<u>Remark 3.9</u> Observe that $y(t)^T f(x_t^o,u^o(t),t)$ is scalar. Hence the second derivative with respect to x_t^o, appearing in (3.16) is a bilinear form in $(\varphi,\psi) \in C(-r,0;R^n)\times C(-r,0;R^n)$.

By an extension of the Riesz representation theorem, such bilinear forms can be represented as repeated Riemann Stieltjes integrals,

Fréchet [1915], yielding

$$\int_{-r}^{0} \varphi(s)^T d_s \int_{-r}^{0} d_\tau K(s,\tau)\psi(\tau),$$

where $K(s,\tau)$ is a $n\times n$-matrix function; each component of K has a finite F-variation on $[-r,0]\times[-r,0]$ (compare also Morse [1950]).

CHAPTER VI

LOCAL RELAXED MINIMA

In this chapter, we discuss the relaxed version of the optimal control problem introduced in the preceeding chapter. We employ J. Warga's formulation of relaxed problems and refer for many technical and heuristic details to Warga's book [1972; for the latter to Chapter III].

Section 2 clarifies the relation to the ordinary problem, in particular, it turns out that for periodic problems "frequently" every *ordinary* optimal solution is also optimal among *relaxed* solutions.

Then the development follows closely that of Chapter V. In Sections 3 and 4, first and second order necessary optimality conditions are proven, based on the Banach space methods of Sections II.1 and II.2 (for simplicity, we omit the state constraint). Many proofs are simple analogues of those in Chapter V and hence omitted. Hopefully, this will emphasize the main difference: First order conditions for the relaxed problem have the form of a *global* maximum principle as in Chapter IV. The second order conditions are new. At first glance applicability of the Banach space methods may appear strange, since strong variations, on which the proof in Chapter IV was based, are not even Gateaux derivatives. However, relaxed controls are equipped with other linear and topological structures than ordinary controls, allowing to obtain the desired Fréchet differentiability properties.

The time interval T is kept fixed throughout.
The main results of this chapter are Theorem 2.5, Corollary 3.3 and Theorem 4.4.

1. Problem Formulation

We start defining relaxed controls, and give a brief discussion of their properties, following Warga [1972].

Suppose that

$$\Omega(t) \subset \Omega_0 \quad \text{for a.a.} \quad t \in T,$$

where $\Omega_0 \subset R^m$ is compact, $t \to \Omega(t)$ is measurable and $\Omega(t)$ is closed for a.a. $t \in T$.

The set of Radon probability measures on Ω_0 is denoted by $rpm(\Omega_0)$. Now consider the Banach space

$$L^1(T,C(\Omega_0)).$$

The dual space $N = (L_1(T,C(\Omega_0))^*$ can be identified with the space of all (equivalence classes) of weak* measurable functions

$$v:T \to C(\Omega_0)^*,$$

with

$$|v|:= \operatorname{ess\,sup}_T |v(t)| = \operatorname{ess\,sup}_T \{ \sup_{|\varphi|_\infty \le 1} \int_\Omega \varphi(\omega)v(t)(d\omega)\} < \infty.$$

This is proven in Warga [1972,Theorem IV.1.8] and one can also deduce it from a general theorem due to Dinculeanu (cp. Diestel/Uhl [1977]).

A relaxed control $v \in S^\#$ is an element of N with values $v(t) \in rpm(\Omega_0)$ having support contained in $\Omega(t)$. In the case $\Omega(t) \equiv \Omega_0$, we simply write $v \in S$. Thus relaxed controls satisfy the following measurability property:

$$t \to c(v(t)):= \int_{\Omega_0} c(\omega)v(t)(d\omega)$$

is measurable for each $c \in C(\Omega_0)$.

For a function $f: C(-r,0;R^n)\times R^m \times T \to R^n$ which is jointly continuous in the first two arguments and measurable in the third one, we define

$$f(\varphi,v(t),t):= \int_{\Omega_0} f(\varphi,\omega,t)v(t)(d\omega);$$

this is measurable in t by weak* measurability of v. An ordinary control $u(\cdot):T \to \Omega_0$ is identified with the relaxed control $v(\cdot) = \delta_{u(\cdot)}$, where $\delta_{u(t)}$ denotes the point measure concentrated at $u(t) \in \Omega_0$.

The main advantage of relaxed controls compared to ordinary controls is that the set $S^\#$ equipped with the weak* topology is compact and sequentially compact, while ordinary controls are dense in $S^\#$. This facilitates very much proof of existence for (approximate) optimal controls; sometimes, only existence of a relaxed optimal solution as (weak*) limit in numerical procedures can be established, Williamson/Polak [1976].

We formulate the following relaxed problems.

<u>Problem 1.1</u> Minimize $\int_{t_1}^{t_2} g(x(t),v(t),t)dt$

s.t. $\dot{x}(t) = f(x_t,v(t),t)$ a.a. $t \in T := [t_1,t_2]$

$p(x_{t_2},x_{t_1}) = 0$

$v \in S^{\#}$

where f,g and p are as in Problem V.1.1.

<u>Problem 1.2</u> Minimize $1/(t_2-t_1) \int_{t_1}^{t_2} g(x(t),v(t),t)dt$

s.t. $\dot{x}(t) = f(x_t,v(t),t)$ a.a. $t \in T := [t_1,t_2]$

$x_{t_1} = x_{t_2}$

$v \in S^{\#}$.

We will prove necessary optimality conditions for pairs (x^0,v^0) which are optimal in the following sense.

<u>Definition 1.3</u> A pair $(x^0,v^0) \in C(t_1-r,t_2;R^n)\times S^{\#}$ satisfying the constraints of Problem 1.1 is called a *local relaxed minimum* if for some $\varepsilon > 0$ all such pairs (x,v) with

$$|x-x^0|_\infty < \varepsilon$$

satisfy

$$\int_{t_1}^{t_2} g(x^0(t),v^0(t),t)dt \le \int_{t_1}^{t_2} g(x(t),v(t),t)dt.$$

Henceforth, we assume that (x^0,v^0) is a local relaxed minimum of Problem 1.1 and define

$$\varphi^0 := x^0_{t_1}.$$

<u>Remark 1.4</u> If v^0 coincides with an ordinary control u^o, i.e. $v^0(\cdot) = \delta_{u^o(\cdot)}$, then (x^0,u^o) is a strong local minimum.

The converse is in general false; but compare section 2, below.

<u>Remark 1.5</u> It does not make much sense (though it is possible) to distinguish between weak and strong local relaxed minima, since in every neighborhood (with respect to the norm topology) of $v^0 \in S^{\#}$ there exists $v \in S^{\#}$ such that for a.a. $t \in T$, $v(t)$ has support

on all of $\Omega(t)$.

The following assumptions on the data of Problem 1.1 will be imposed throughout this chapter, mostly without further mentioning.

Hypothesis 1.6 The functions $g(x,\omega,t)$ and $f(\varphi,\omega,t)$ are continuous in (x,ω) and (φ,ω), respectively, and measurable in t. They are Fréchet differentiable with respect to their first arguments, and the derivatives $g_x(x,\omega,t)$ and $\mathcal{D}_1 f(\varphi,\omega,t)$ are jointly continuous in (x,ω) and (φ,ω), respectively. The function p is continuously Fréchet differentiable.

Hypothesis 1.7 There is $q: R_+ \times R \to R_+$ such that for all $x \in R^n$, $\varphi \in C(-r,0;R^n)$ and $\omega \in \Omega_0$

$$|g(x,\omega,t)| + |g_x(x,\omega,t)| \le q(|x|,t),$$

$$|f(\varphi,\omega,t)| + |\mathcal{D}_1 f(\varphi,\omega,t)| \le q(|\varphi|,t) \quad \text{for a.a. } t \in T$$

where $q(s,\cdot) \in L^2(T;R)$ for all $s \in R_+$ and $q(\cdot,t)$ is monotonically increasing for a.a. $t \in T$.

Hypothesis 1.8 The initial value problem

$$x_{t_1} = \varphi^0, \dot{x}(t) = f(x_t, v(t), t), \quad \text{a.a. } t \in T$$

has for all $v \in S^*$ a solution (uniqueness will follow from the other assumptions).

We note some consequences of these assumptions, without giving proofs (see also Colonius [1982a]).

Lemma 1.9 Let $(x^k, v^k) \subset C(t_1-r, t_2; R^n) \times S^*$ be a sequence with $x^k \to x^0$ uniformly and $v^k \to v^0$ weakly* in S^*. Then

$$(f(x_t^k, v^k(t), t), t \in T) \to (f(x_t^0, v^0(t), t), \ t \in T)$$

weakly in $L^2(T;R^n)$.

Variants of this lemma are well-known, even for functions f which allow lags in the control, Berkovitz [1975], Warga [1972,1974], Bates [1977].

Lemma 1.10 The Fréchet derivative $\mathcal{D}_1 f(\varphi, v(t), t)$ exists and has the form

$$\mathcal{D}_1 f(\varphi, v(t), t)\psi = \int_{\Omega_0} \mathcal{D}_1 f(\varphi,\omega,t)\psi v(t)(d\omega).$$

Lemma 1.11 Let $((x^k, v^k)) \subset C^n(t_1-r, t_2; R^n) \times S^{\#}$ converge in the norm to (x^0, v^0). Then

$$\operatorname{ess\,sup}_t |f(x_t^k, v^k(t), t) - f(x_t^o, v^0(t), t)| \to 0$$

and

$$(|D_1 f(x_t^k, v^k(t), t) - D_1 f(x_t^o, v^0(t), t)|,\ t \in T) \to 0$$

weakly in $L^2(T; R^n)$.

The following lemma allows us to characterize relaxed velocity vectors.

Lemma 1.12 For a measurable subset $S \subset T$, consider a measurable function $z: S \to R^n$ and a function $\Phi: \Omega_0 \times S \to R^n$ with $\Phi(\omega, \cdot)$ measurable for all $\omega \in \Omega_0$ and $\Phi(\cdot, t)$ continuous for a.a. $t \in S$. Then the following three conditions are equivalent:

(i) $z(t) \in \operatorname{co}\Phi(\Omega(t), t)$ a.a. $t \in S$;

(ii) $z(t) = \Phi(v(t), t)$ a.a. $t \in S$

for an element $v \in S^{\#}$;

(iii) $z(t) = \sum_{i=0}^{n} \alpha_i(t)\Phi(u_i(t), t)$ a.a. $t \in S$

for some measurable $\alpha_i: T \to R_+$ and $u_i: T \to \Omega_0$ with $\sum_{i=0}^{n} \alpha_i(t) = 1$ and $u_i(t) \in \Omega(t)$.

Proof: The proof follows by Warga [1972, Theorems I.6.13, IV.3.13; compare also Theorem VI.3.2]. □

Remark 1.13 Consider

$$\{(f(x_t, v(t), t), t \in T): v \in S^{\#}\}.$$

Then the lemma above implies that this set coincides with

$$\{z \in L_\infty^n(T): z(t) \in \operatorname{co} f(x_t, \Omega(t), t) \text{ a.a. } t \in T\}.$$

This shows that along a fixed trajectory x the set of relaxed velocity vectors coincides with the convex hull of the set of ordinary velocity vectors. Hence the relaxed system is equivalent to the relaxed system considered by Oguztöreli [1966, §8.9].

Remark 1.14 The existence and uniqueness property required in Hypothesis 1.8 for trajectories corresponding to relaxed controls $v \in S^{\#}$ can be reduced to existence and uniqueness theory of functional differential equations using the representation of relaxed trajectories introduced by Gamkrelidze: by Lemma 1.12, for each relaxed trajectory there exist measurable functions $\alpha_0,\alpha_1,\ldots,\alpha_n:T \to R_+$ with $\Sigma\, \alpha_i(t) = 1$ and ordinary controls $u_0,u_1,\ldots,u_n$ with values in $\Omega(t)$ such that

$$\dot{x}(t) = \sum_{i=0}^{n} \alpha_i(t) f(x_t,u_i(t),t) \quad \text{a.a.} \quad t \in T,$$

and conversely. Hence relaxed trajectories (i.e. trajectories corresponding to a relaxed control $v \in S^{\#}$) satisfy a functional differential equation.

Now we are in a position to develop the theory for the relaxed problem in analogy to Chapter V. There exists an open set $\mathcal{O} \subset C(T;R^n)\times N\times C(-r,0;R^n)$ containing (x^0,v^0,φ^0) such that the following maps G and F with values in R and $C(T;R^n)$ respectively, are well-defined on $\mathcal{O}$:

$$G(x,v) := \int_{t_1}^{t_2} g(x(t),v(t),t)\,dt \tag{1.1}$$

and

$$[F(x,v,\varphi)](t) := \varphi(0) + \int_{t_1}^{t} f(x_s,v(s),s)\,ds, \quad t \in T; \tag{1.2}$$

here it is understood that at the right hand side

$$x(t_1+s) := \varphi(s), \quad s \in [-r,0].$$

For simplicity we take $\mathcal{O} = \tilde{\mathcal{O}}\times N\times \mathcal{O}_\varphi$ such that for all $x \in \tilde{\mathcal{O}}$ one has $x_t \in \mathcal{O}_\varphi$ for all $t \in T$.

Note that the maps F and G are bounded and linear in $v \in N$; furthermore, they are also continuous in v with respect to the weak* topology on N.

We get the following analogue of Lemma V.1.11.

Lemma 1.15 (a) The map F is continuously Frêchet differentiable and

$$Id - \mathcal{D}_1F(x^0,v^0,\varphi^0)$$

is an isomorphism on $C(T;R^n)$.

(b) The linearized equation

$$x = \mathcal{D}F(x^0,v^0,\varphi^0)(x,v,\varphi), \quad v,v^0 \in N \tag{1.3}$$

is equivalent to

$$\dot{x}(t) = \mathcal{D}_1 f(x_t^o, v^0(t), t) x_t + f(x_t^o, v(t), t), \quad \text{a.a. } t \in T \tag{1.4}$$

$$x_{t_1} = \varphi.$$

(c) Equation (1.3) has for every $(v,\varphi) \in N \times O_\varphi$ a unique solution x; the equation

$$x = F(x,v,\varphi) \tag{1.5}$$

has for every $v \in S^{\#}$, $\varphi \in O_\varphi$ a unique solution $x(v,\varphi)$.

(d) The solution map $S: S^{\#} \times O_\varphi \to C(T;R^n)$ of (1.3) defined as

$$S(v,\varphi) := x(v,\varphi),$$

is continuously Fréchet differentiable, and

$$x := \mathcal{D}S(v^0,\varphi^0)(v,\varphi)$$

is the solution of (1.4).

Define

$$P(x,\varphi) = p(x_{t_2},\varphi) \tag{1.6}$$

where $x(t_1+s) = \varphi(s)$, $s \in [-r,0]$, if $t_2 < t_1+r$.

Now Problem 1.1 can be reformulated as

Problem 1.16 Minimize $G(S(v,\varphi),\varphi)$

over all $(v,\varphi) \in N \times O_\varphi$

satisfying $P(S(v,\varphi),\varphi) = 0$

$v \in S^{\#}$.

We note the following result.

Lemma 1.17 If (x^0,v^0) is a local relaxed minimum of Problem 1.1, then $(v^0,x^o_{t_1})$ is a local minimum of Problem 1.16.

2. Relations between Ordinary and Relaxed Problems

Here we analyze the relations between the relaxed periodic problem 1.2 and the associated "ordinary" problem, where instead of relaxed controls $v \in S^{\#}$ measurable control functions $u: T \to \Omega_o$ are considered.

Thus we are interested in pairs $(x,u) \in C(t_1-r,t_2,R^n) \times L^\infty(t_1,t_2;R^m)$ s.t.

$$\dot{x}(t) = f(x_t, u(t), t), \quad \text{a.a.} \quad t \in T \tag{2.1}$$

$$x_{t_1} = x_{t_2} \tag{2.2}$$

$$u \in \mathcal{U}_{ad}(T) := \{u \in L^\infty(T;R^m) : u(t) \in \Omega(t) \text{ a.e.}\}. \tag{2.3}$$

Recall that every $u \in \mathcal{U}_{ad}(T)$ can be identified with

$$v(\cdot) := \delta_{u(\cdot)} \in S^{\#}.$$

The following hypothesis is used throughout this section.

Hypothesis 2.1 There exists a bounded set $B \subset C(t_1-r, t_2; R^n)$ such that for every $v \in S^{\#}$, the equation

$$\dot{x}(t) = f(x_t, v(t), t), \quad \text{a.a.} \quad t \in T \tag{2.4}$$

has a solution $x = x(v) \in B$ with $x_{t_1} = x_{t_2}$.

Remark 2.2 Sufficient conditions for the property above can be given based on Remark 1.14 and using e.g. assumptions and methods similar to Nistri [1983]; cp. also Russell [1982]. We do not go into this vast field here, and are content with stating the property which is needed in the sequel.

Lemma 2.3 Let (v^n) be a sequence in $S^{\#}$ converging in the weak* topology to $v^o \in S^{\#}$. Then a subsequence of $x^n := x(v^n) \in B \subset C(t_1-r, t_2; R^n)$ satisfying

$$x^n_{t_1} = x^n_{t_2}, \quad \dot{x}^n(t) = f(x^n_t, v^n(t), t), \quad \text{a.a.} \quad t \in T \tag{2.5}$$

converges uniformly to a solution x^o of

$$x^o_{t_1} = x^o_{t_2}, \quad \dot{x}^o(t) = f(x^o_t, v^o(t), t), \quad \text{a.a.} \quad t \in T.$$

Proof: By Hypothesis 2.1, existence of periodic $(x^n) \subset B$ with $\dot{x}^n(t) = f(x^n_t, v^n(t), t)$ for a.a. $t \in T$ is guaranteed. Furthermore, (x^n) is equicontinuous, since

$$|x^n(t)-x^n(t')| \le \int_{t'}^{t} |f(x^n_s, v^n(s), s)|ds \le \int_{t'}^{t} \max_{\omega \in \Omega_o} |f(x^n_s, \omega, s)|ds$$
$$\le \int_{t'}^{t} q(c^o, s)ds$$

where q is given by Hypothesis 1.7 and c^o is a constant determined by B.

Thus by Arzèla-Ascoli's Theorem, a subsequence (x^{n_k}) converges uniformly to $x^o \in \mathcal{B}$.

By Lemma 1.9

$$(f(x_t^{n_k},v^{n_k}(t),t) - f(x_t^o,v^o(t),t),\ t \in T) \to 0 \quad \text{weakly in} \quad L_2(T;R^n).$$

Thus for all $t \in T$

$$\begin{aligned} x^o(t) &= x^o(t_1) + \lim_{k\to\infty} \int_{t_1}^{t} f(x_s^{n_k},v^{n_k}(s),s)ds \\ &= x^o(t_1) + \int_{t_1}^{t} f(x_s^o,v^o(s),s)ds. \end{aligned}$$

and, naturally,

$$x_{t_1}^o = x_{t_2}^o.$$

□

We have the following easy consequence on existence of a relaxed minimum.

Theorem 2.4 Suppose that Problem 1.2 satisfies Hypotheses 1.6, 1.7, and 2.1. Then there exists a relaxed minimum (x^o,v^o).

Proof: This follows by sequential compactness of S^*, Warga [1972, Theorem IV.3.11], Lemmas 2.3 and 1.9. □

By Warga [1972,Theorem IV.3.10], ordinary controls are dense in S^*. Thus one will expect that every relaxed trajectory x can be uniformly approximated by ordinary trajectories, which will *approximately* satisfy the boundary condition ("minimizing approximate $\mathcal{U}$-solutions" in Warga's terminology). However, more is true under Hypothesis 2.1: Every optimal relaxed solution can be approximated by ordinary trajectories which *satisfy* the boundary condition.

Theorem 2.5 Suppose that Problem 1.2 satisfies Hypotheses 1.6, 1.7, and 2.1. If in Hypothesis 2.1 $x(v)$ is uniquely determined by v, then

$$\inf 1/(t_1-t_2) \int_{t_1}^{t_2} g(x(t),v(t),t)dt = \inf 1/(t_2-t_1) \int_{t_1}^{t_2} g(x(t),u(t),t)dt$$

where the infimum at the left hand side is taken over all pairs (x,v) satisfying the constraints of Problem 1.2, and the infimum at the right hand side is taken over all pairs (x,u) satisfying (2.1) - (2.3).

Proof: By Theorem 2.4 the infimum at the left hand side is actually attained, say by (x^o,v^o). By Lemma 2.3 and weak* density of ordinary controls in $S^\#$, Warga [1972, Theorem IV.3.10], there exist sequences $v^n(\cdot) = \delta_{u^n(\cdot)} \in S^\#$ and $(x^n) \subset C(t_1-r,t_2;R^n)$ with $v^n \to v^o$ and $x^n \to x^o$ satisfying (2.1) - (2.3). By Lemma 1.9,

$$\lim_{n\to\infty} 1/(t_2-t_1) \int_{t_1}^{t_2} g(x^n(t),u^n(t),t)dt = 1/(t_2-t_1) \int_{t_1}^{t_2} g(x^o(t),v^o(t),t)dt$$

□

Remark 2.6 In particular, under the assumptions of Theorem 2.5, every strong local minimum is a local relaxed minimum. This result is a crude reflection of Warga's beautiful [1971] result on abnormality in problems with optimal ordinary solutions which are not optimal among relaxed solutions. Hypothesis 2.1 excludes, in some sense, the abnormal case.

Observe that, naturally, Theorem 2.5 allows for relaxed minima which are not ordinary solutions. This is important in aircraft flight performance optimization ("chattering cruise", Speyer [1973], Houlihan/Cliff/Kelley [1982]), and in certain control problems for chemical reactors (see e.g. Horn/Bailey [1968]).

Example 2.7 Suppose the system equation is given by

$$\dot{x}(t) = L(t,x_t) + b(v(t)), \quad t \in T$$

with L as in (III.1.2) and

$$b \in C(\Omega;R^n), \quad \Omega \subset R^m \text{ compact.}$$

Assume that

$$\mathrm{Ker}[\Phi(t_2,t_1)-\mathrm{Id}] = \{0\}$$

where $\Phi(t,s)$ is the associated family of evolution operators.

Then the strengthened Hypothesis 2.1 is satisfied, and every ordinary optimal control is also a relaxed optimal control.

Note that this is true without any convexity condition for

$$\left\{ \begin{pmatrix} g(x(t),\omega,t) \\ L(t,x_t) + b(\omega,t) \end{pmatrix} : \quad \omega \in \Omega \right\}.$$

This result is reminiscent of Neustadt [1963]. However the arguments here are entirely different. Note that Neustadt's arguments cannot be extended to functional differential equations with function space boundary conditions, since they are based on the bang-bang principle.

3. First Order Necessary Optimality Conditions

We prove first order necessary optimality conditions for the relaxed Problems 1.1 and 1.2 and discuss the required constraint qualification. Furthermore the fixed boundary value problem is considered.

Derivatives of $S(v,\varphi^0)$ at $v = v^0$ in direction $\alpha(v-v^0) \in S^{\#}(v^0)$ will be needed. Note that $v-v^0 \notin S^{\#}$ for $v,v^0 \in S^{\#}$. Thus it is convenient to write the linearized equation

$$x = \mathcal{D}_1 S(v^0,\varphi^0)(\alpha(v-v^0)), \quad \alpha \in R_+, \quad v \in S^{\#} \tag{3.1}$$

in the form

$$x_{t_1} = 0, \; \dot{x}(t) = \mathcal{D}_1 f(x_t^0, v^0(t),t)x_t + u(t), \quad \text{a.a.} \quad t \in T \tag{3.2}$$

with $u \in V_{ad}(v^0) := \{u \in L^\infty(T;R^n) : u(t) \in K(t) \text{ a.e.}\}$ where $K(t) \subset R^n$ is the closed and convex cone defined by

$$K(t) := R_+[\text{co}f(x_t^0,\Omega(t),t) - f(x_t^0,v^0(t),t)], \quad t \in T.$$

The hypotheses and Lemma 1.15 assure the required differentiability properties of G and S.

Define the attainability cone A for Problem 1.1 by

$$\begin{aligned} A = \{\mathcal{D}p(x_{t_2}^0,\varphi^0)(x_{t_2},\varphi) &: \varphi \in C(-r,0;R^n) \text{ and} \\ &\text{there exists } u \in V_{ad}(v^0) \text{ s.t. } x_{t_1} = \varphi \\ &\text{and } \dot{x}(t) = \mathcal{D}_1 f(x_t^0,v^0(t),t)x_t + u(t), \text{ a.a. } t \in T\}. \end{aligned} \tag{3.3}$$

Lemma 1.15 and Remark 1.13 imply

<u>Lemma 3.1</u> The cone A defined above satisfies

$$A = \{\mathcal{D}P(x^0,\varphi^0)(\mathcal{D}S(v^0,\varphi^0)(v,\varphi),\varphi) : \varphi \in C(-r,0;R^n), \; v \in S^{\#}(v^0)\}.$$

Define the Lagrangean L for Problem 1.16 as

$$L(v,\varphi,\lambda) := \lambda_0 G(S(v,\varphi),v) - y^*P(S(v,\varphi),\varphi),$$

where $(v,\varphi) \in N \times O_\varphi$ and $\lambda = (\lambda_0,y^*) \in R \times Y^*$.

The following first order necessary optimality conditions for

Problem 1.16 hold.

Theorem 3.2 Let $(v^0,\varphi^0) \in N\times O_\varphi$ be a local minimum of Problem 1.16, define $x^0 := S(v^0,\varphi^0)$ and assume that the attainability cone A defined in (3.3) contains a subspace of finite codimension in Y.

Then there exists $0 \neq \lambda = (\lambda_0, y^*) \in R_+\times Y^*$ such that

$$\mathcal{D}_1 L(v^0,\varphi^0)v \geq 0 \quad \text{for all} \quad v \in S^{\#}(v^0) \tag{3.5}$$

$$\mathcal{D}_2 L(v^0,\varphi^0) = 0 \quad \text{in} \quad C(-r,0;R^n)^*. \tag{3.6}$$

If the constraint qualification $A = Y$ satisfied then $\lambda_0 \neq 0$.

Proof: Follows as Theorem V.2.4 from Theorem II.1.11. □

We proceed to the periodic Problem 1.2. Here the attainability cone A has the form

$$\begin{aligned} A = \{x_{t_2} - \varphi : \varphi \in C(-r,0;R^n) \\ \text{and there exists} \quad u \in V_{ad}(v^0) \quad \text{s.t.} \quad x_{t_1} = \varphi \\ \text{and} \quad \dot{x}(t) = \mathcal{D}_1 f(x_t^o, v^0)(t),t)x_t + u(t) \text{ a.a. } t \in T\}. \end{aligned} \tag{3.7}$$

We obtain the following first order necessary optimality conditions for the periodic Problem 1.2, which have the form of a global maximum principle.

Corollary 3.3 Let $t_2 \geq t_1 + r$ and suppose that $(x^0,v^0) \in C(t_1-r,t_2;R^n) \times S^{\#}$ is a local relaxed minimum of Problem 1.2, where Hypotheses 1.6 - 1.8 are satisfied. Then there exist $\lambda_0 \geq 0$ and a (t_2-t_1)-periodic solution y of the adjoint equation

$$\begin{aligned} \frac{d}{ds}\{y(s)-y(t_2) + \int_s^{t_2+r} [\eta^T(\alpha,s-\alpha) - \eta^T(\alpha,t_2-\alpha)]y(\alpha)d\alpha\} \\ = -\lambda_o g_x(x^0(s)u^o(s),s), \quad \text{a.a.} \quad s \leq t_2 \end{aligned} \tag{3.8}$$

such that $(\lambda_0, y^{t_2}) = \lambda \neq 0$ in $R_+\times NBV(0,r;R^n)$ and

$$\begin{aligned} \lambda_o g(x^0(s),v^0(s),s) + y(s)^T f(x_s^o, v^0(s),s) \\ = \min_{\omega\in\Omega(s)} \{\lambda_0 g(x^0(s),\omega,s) + y(s)^T f(x_s^o,\omega,s)\} \text{ for a.a. } t \in T. \end{aligned} \tag{3.9}$$

If the attainability cone A specified in (3.7) satisfies $A = C(-r,0;R^n)$, then $\lambda_0 \neq 0$.

Here η is given by the representation

$$\mathcal{D}_1 f(x_s^o, v^0(s), s)\varphi = \int_{-r}^{0} [d_t\eta(s,t)]\varphi(t), \quad \varphi \in C(-r,0;R^n).$$

Proof: Follows similarly as Theorem V.2.9 from Theorem 3.2. □

Remark 3.4 Note that we do not have to differentiate with respect to ω; the maps F and G are linear in the generalized control v.

Remark 3.5 If v^0 happens to coincide with an ordinary control, i.e. $v^0(\cdot) = \delta_{u^o(\cdot)}$, then the assertions of this theorem reduce to those of the global maximum principle Theorem IV.2.1. Note, however, that - in the absence of Hypothesis 2.1 - even in this case, the optimality requirements in Corollary 3.3 are stronger, since we require that (x^0, v^0) is optimal among relaxed solutions (x,v).

Remark 3.6 Suppose that f and g are affinely linear in ω with $\Omega(t)$ convex for a.a. $t \in T$. Then by Remark 1.13, every relaxed trajectory is also an ordinary trajectory, and the optimal values coincide. Hence every strong local minimum is also a local relaxed minimum, and thus the necessary optimality conditions of Corollary 3.3 apply.

We now take up again the discussion of fixed boundary value problems, which we have begun at the end of Section V.2. Here

$$p(x_{t_2}, x_{t_1}) = (x_{t_2} - \varphi^2,\; x_{t_1} - \varphi^1) \tag{3.11}$$

with $\varphi^1, \varphi^2 \in C(-r,0;R^n)$ fixed. For simplicity we assume

$$\Omega(t) \equiv \Omega_o \quad \text{and } f \text{ is independent of } t. \tag{3.12}$$

Define for an interval $I \subset T$ and $p = 2$ or $p = \infty$

$$U^p(I) := \{u \in L^p(I;R^n) : u(t) \in K(t) \text{ a.e.}\} \quad \text{where}$$

$$K(t) := R_+[\operatorname{cof}(x_t^o, \Omega) - f(x_t^o, v^0(t))]$$

and consider

$$x_{t_1} = 0, \; \dot{x}(t) = \mathcal{D}_1 f(x_t^o, v^0(t)x_t + u(t), \text{ a.a. } t \in T. \tag{3.13}$$

Define for $p = 2$ and $p = \infty$ the cone (cp.(V.2.29))

$$A_o^p := \{x_{t_2} : \text{there exists } u \in L^p(T;R^n) \text{ s.t. } x \text{ solves (3.13)}\}. \tag{3.14}$$

For an application of Theorem 3.2 we have to analyse if

$$A_o^\infty \text{ has finite codimension in } Y := W_\infty^{(1)}(-r,0;R^n). \tag{3.15}$$

We may disregard the R^n-component of x_{t_1}.

Now, by Colonius [1982a,Theorem3.3] condition (3.15) implies that

$$\operatorname{int} \{z \in L^\infty(t_2-r,t_2;R^n): z(t) \in K(t) \text{ a.e.}\} \neq \emptyset. \quad (3.16)$$

Example 3.1 in the cited paper shows that zero may be a boundary point of $K(t)$ on a set of points t of non zero measure, while (3.15) holds.

On the other hand, the following regularity condition

$$\dot{\varphi}^2(t-t_2) = f(x_t^0,v^0(t)) \in \operatorname{int}_\delta \operatorname{cof}(x_t^0,\Omega) \text{ a.a. } t \in [t_2-r,t_2] \quad (3.17)$$

where $\operatorname{int}_\delta A$ for $A \subseteq R^n$ denotes the set of all elements in A which have at least distance $\delta > 0$ to the boundary of A, implies (3.15).

Obviously, if $f(\varphi,\omega)$ is nonlinear in ω, the condition $\operatorname{int} \operatorname{cof}(x_t^0,\Omega) \neq \emptyset$ and hence (3.17) may be satisfied also for scalar controls ω.

Thus for the relaxed problem, the analogue of the rank condition (V.2.30), is much less restrictive.

By Theorem 3.2 condition (3.15) implies the existence of nontrivial Lagrange multipliers $(\lambda_0,y^*) \in R \times W_\infty^{(1)}(-r,0;R^n)^*$. Since $W_\infty^{(1)}(-r,0;R^n)^*$ is not identifiable with a space of real-valued functions on $[-r,0]$, it is very important to get more regularity properties of y^*. This can be accomplished using Theorem II.1.18.

We have the following situation:

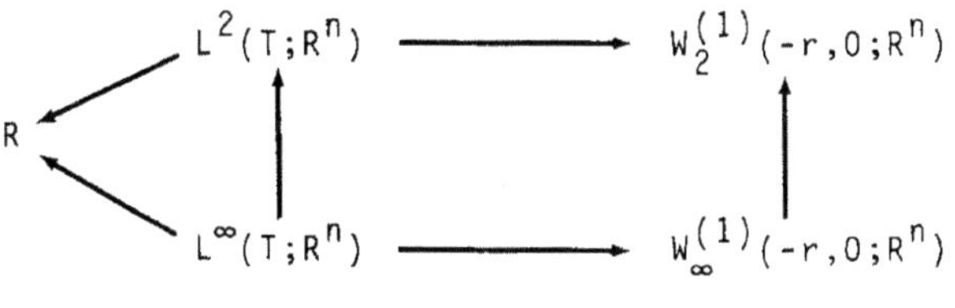

Figure 2

We cite the following result, Colonius [1982a,Addendum Theorem 1].

<u>Proposition 3.7</u> The following two conditions are equivalent:

(i) $\{(\dot{x})_{t_2}: x \text{ solves (3.13) for some } u \in U^2(T)\} = L^2(-r,0;R^n)$

(ii) $K(t) = R^n$ for a.a. $t \in [t_2-r,t_2]$.

Furthermore, if for a.a. $t \in [t_2-r,t_2]$, the cone $K(t)$ equals R^n, then the following two conditions are equivalent:

(iii) $U^\infty([t_2-r,t_2]) = \{\alpha u : \alpha \in R_+,\ u \in L^\infty(t_2-r,t_2;R^n)$ with $u(t) \in \mathrm{cof}(x_t^0,\Omega) - f(x_t^0,v^0(t))$ a.e.$\}$.

(iv) For some $\delta > 0$

$$0 \in \mathrm{rel\ int}_\delta\ \mathrm{cof}(x_t^0,\Omega) - f(x_t^0,v^0(t)) \text{ for a.a. } t \in [t_2-r,t_2];$$

here rel int_δ A for $A \subset R^n$ denotes the set of all elements in A which have at least distance δ to the boundary of A with respect to the smallest linear subspace containing A.

The set at the left hand side of the equation in (i) does not change, if $u|[t_1,t_2-r]$ is required to lie in L^∞. The cone

$$\{\alpha u : \alpha \in R_+,\ u \in L^\infty(T;R^n) \text{ with } u(t) \in \mathrm{cof}(x_t^0,\Omega) - f(x_t^0,v^0(t)) \text{ a.e.}\}$$

corresponds to the cone of admissible directions for the control constraint.

But, by Example II.1.20, the L^2-closure of $U^\infty([t_2-r,t_2])$ contains $U^2([t_2-r,t_2])$. Hence, taken together, the regularity condition (3.17) (being equivalent to (ii),(iv)), means by (i) and (iii) that the L^2-closure of the cone of admissible directions is mapped onto $L^2(-r,0;R^n)$ under the linearized control-to-final-state-velocity map. If this holds, the assumptions of Theorem II.1.18 are satisfied and a Lagrange multiplier in $W_\infty^{(1)}(-r,0;R^n)^*$ can be identified with an element in $W_2^{(1)}(-r,0;R^n)^*$ (observe that the finite dimensional part $x(t_2-r) = \varphi^2(-r)$ does not pose any problem here).

Furthermore, Proposition 3.7 shows very clearly, where the uniformity condition (that is the δ-bound) in (3.17) comes in: It guarantees that the cone U^∞ defined by pointwise restrictions is not larger than the cone of admissible directions (on the relevant interval $[t_2-r,t_2]$).

Condition (3.17) is used for more general boundary conditions in Colonius [1982a,b].

The following theorem (the proof of which will be omitted) indicates the concrete form of the necessary optimality conditions obtained under the assumption (3.17). Note that, by duality theory of functional differential equations, the dual space of $W_2^{(1)}(-r,0;R^n)$ has to be identified with $R^n \times L^2(-r,0;R^n)$, cp. also Section III.2. The adjoint equation has an integrated form.

Theorem 3.8 Let $(x^0,v^0) \in C(t_1-r,t_2) \times S^\#$ be a local relaxed minimum of the fixed boundary value Problem (i.e. Problem 1.1 with p given by (3.11)), satisfying Hypothesis 1.6 - 1.8 and (3.12).

Then there exist $\lambda_0 \geq 0$ and a solution y of the transposed equation on $[t_1,t_2-r]$

$$y(s)-y(t_2-r) = -\int_s^{t_2}[\eta^T(\alpha,s-\alpha) - \eta^T(\alpha,t_2-r-\alpha)]y(\alpha)d\alpha \qquad (3.18)$$

$$+ \lambda_0 \int_s^{t_2} g_x(x^0(t),v^0(t),t)dt$$

such that $(0,0,0) \neq (\lambda_0, y(t_2-r), y^{t_2-r}) \in R \times R^n \times L^2(0,r;R^n)$ and

$$\lambda_0 g(x^0(s),v^0(s),s) + y(s)^T f(x_s^o,v^0(s),s) \qquad (3.19)$$

$$= \min_{\omega\in\Omega} \{\lambda_0 g(x^0(s),\omega,s) + y(s)^T f(x_s^o,\omega,s)\} \text{ for a.a. } t \in T.$$

Remark 3.9 For results related to the boundary value problem occuring in the optimality conditions above see e.g. Hutson [1977] and Kamenskii/Myshkis [1972]. Numerical methods are developed in Wierzbicki/Hatko [1973], Reddien/Trevis [1974], Mathis/Reddien [1978], Bader [1983].

4. Second Order Necessary Optimality Conditions

This section presents second order conditions which are obtained by an application of Corollary II.2.12.

The following hypotheses are imposed throughout this section. They guarantee the required twice continuous Fréchet differentiability of F, G, and S.

Hypothesis 4.1 The functions $g(x,\omega,t)$ and $f(\varphi,\omega,t)$ are twice continuously Fréchet differentiable in x and φ, respectively.

Hypothesis 4.2 For all $x \in \mathcal{O}_x$, $\varphi \in \mathcal{O}_\varphi$ and $\omega \in \Omega_o$

$$|g_{xx}(x,\omega,t)| \leq q(|x|,t) \text{ for a.a. } t \in T$$

$$|\mathcal{D}_1\mathcal{D}_1 f(\varphi,\omega,t)| \leq q(|\varphi|,t) \text{ for a.a. } t \in T,$$

with q as in Hypothesis 1.7.

Defining the Lagrangean

$$L(v,\varphi,\lambda) := \lambda_0 G(S(v,\varphi),v) - y^*F(S(v,\varphi),v) \tag{4.1}$$

where $\lambda = (\lambda_0, y^*) \in R\times C(-r,0;R^n)^*$, we obtain the following abstract second order optimality conditions.

<u>Theorem 4.3</u> Suppose that $(v^0,\varphi^0) \in S^{\#}\times C(-r,0;R^n)$ is a local relaxed minimum of Problem 1.2, and assume that Hypotheses 1.6 - 1.8 and 4.1, 4.2 are satisfied. Define $x^0 := S(v^0,\varphi^0)$ and assume that the attainability cone A defined in (3.3) contains a subspace of finite codimension in Y.

Then for every pair $(v,\varphi) \in N\times C(-r,0;R^n)$ with

$$\mathcal{D}G(x^0,v^0)(x,v) \le 0, \quad \mathcal{D}P(x^0,\varphi^0)(x,\varphi) = 0 \quad \text{and} \quad v \in S^{\#}(v^0) \tag{4.2}$$

where $x := \mathcal{D}S(v^0,\varphi^0)(v,\varphi)$, there exist $0 \neq \lambda = (\lambda_0,y^*) \in R_+\times C(-r,0;R^n)^*$ such that

$$\mathcal{D}_1 L(v^0,\varphi^0,\lambda)v' \ge 0 \quad \text{for all} \quad v' \in S^{\#}(v^0) \tag{4.3}$$

$$\mathcal{D}_2 L(v^0,\varphi^0,\lambda) = 0 \quad \text{in} \quad C(-r,0;R^n)^* \tag{4.4}$$

$$\mathcal{D}_{1,2}\mathcal{D}_{1,2} L(v^0,\varphi^0,\lambda)((v,\varphi),(v,\varphi)) \ge 0. \tag{4.5}$$

If the attainability cone satisfies $A = Y$ then $\lambda_0 \neq 0$.

<u>Proof:</u> Follows by Corollary II.2.12 (cp. Theorem V.3.7). □

The concrete form of the second order necessary optimality conditions for the periodic relaxed Problem 1.2 can conveniently be given using the Pontryagin function H. Recall

$$\lambda(t) = (\lambda_0, y(t))$$

$$H(\varphi,\omega,\lambda(t),t) := g(\varphi(0),\omega,t) - y^T(t)f(\varphi,\omega,t).$$

<u>Theorem 4.4</u> Let $t_2 \ge t_1+r$ and suppose that $(x^0,v^0) \in C(t_1-r,t_2;R^n) \times S^{\#}$ is a local relaxed minimum of Problem 1.2 with Hypotheses 1.6 - 1.8 and 4.1, 4.2 holding.

Let $(x,v) \in C(t_1-r,t_2;R^n) \times S^{\#}$ be given with

$$\int_{t_1}^{t_2} [g_x(x^0(t),v^0(t),t)x(t) + g(x^0(t),v(t) - v^0(t),t)]dt \le 0 \tag{4.6}$$

$$x_{t_1} = x_{t_2}, \quad \dot{x}(t) = D_1 f(x_t^o, v^o(t), t)x_t + f(x_t^o, v(t) - v^o(t), t) \qquad (4.7)$$

a.a. $t \in T$.

Then there exist $\lambda_o \geq 0$ and a (t_2-t_1)-periodic solution of the adjoint equation

$$\frac{d}{ds}\{y(s) + \int_s^{t_2+r} [\eta^T(\alpha, s-\alpha) - \eta^T(\alpha, t_2-\alpha)]y(\alpha)d\alpha \qquad (4.8)$$

$$= g_x(x^o(s), v^o(s), s) \quad \text{a.a. } s \in T$$

such that $0 \neq \lambda = (\lambda_o, y^{t_2})$ in $R \times NBV(0,r;R^n)$ and for $\lambda(t) = (\lambda_o, y(t))$

$$H(x_s^o, v^o(s), \lambda(s), s) = \min_{\omega \in \Omega(s)} H(x_s^o, \omega, \lambda(s), s) \text{ a.a. } s \leq t_2. \qquad (4.9)$$

$$\int_{t_1}^{t_2} [D_1 D_1 H(x_s^o, v^o(s), \lambda(s), s)(x_s, x_s) \qquad (4.10)$$

$$+ 2D_1 H(x_s^o, v(s) - v^o(s), \lambda(s), s)x_s]ds \geq 0.$$

Proof: The condition (4.10) follows from (4.5) by observing that L defined in (4.1) is affinely linear in v. □

Remark 4.5 Suppose that f and g are affinely linear in ω, and that $\Omega(s)$ is convex for all $s \in T$. Then condition (4.10) reduces to (V.3.16). Observe that here, by Remarks 1.4 and 1.13, strong local minima are also local relaxed minima.

CHAPTER VII

TESTS FOR LOCAL PROPERNESS

In this chapter, we take up the question formulated in the introduction: Suppose an optimal steady state (x^0,u^0) is given. How can one decide, whether it is also optimal among periodic solutions?

A convenient method is to check whether (x^0,u^0) violates necessary optimality conditions for the periodic problem. If this is the case, (x^0,u^0) clearly cannot be an optimal solution for the latter problem. We will muster our arsenal of necessary conditions developed in Chapters IV - VI in order to see which of them give efficient tests for properness.

Considered at a steady state, the necessary optimality conditions for the periodic problem simplify. Hence they will be restated here.

Section 2 shows that the local minimum principle of Section V.2 and the relaxed minimum principle of Section VI.3 are satisfied by every local minimum of the steady state problem and the relaxed steady state problem, respectively. Furthermore it is analysed when the corresponding sets of Lagrange multipliers coincide.

This is used in Section 3 to prove a so-called Π-Test for local properness based on second order weak variations. Section 4 presents a simplified version of this test applying if local properness occurs for all sufficiently high frequencies. Finally, Section 5 discusses tests based on strong variations.

The main results of this chapter are Corollary 2.4, Proposition 2.7, Theorem 3.3, Theorem 4.4 and Theorem 5.1.

1. Problem Formulation

In this chapter we consider, together with its steady state version, the following *autonomous* optimal periodic control problem.

<u>Problem 1.1</u> Minimize $1/\tau \int_0^\tau g(x(s),u(s))ds$

$$\begin{array}{lll} \text{s.t.} & \dot{x}(t) = f(x_t, u(t)), & \text{a.a. } t \in T := [0,\tau] \\ & h(x(t)) \in R^{\ell}_{-} & \text{all } t \in T \\ & \int_0^{\tau} k(x(t),u(t))dt = 0 & \\ & u(t) \in \Omega & \text{a.a. } t \in T \\ & x_0 = x_{\tau} \ ; & \end{array}$$

here $g: R^n \times R^m \to R$, $f: C(-r,0;R^n) \times R^m \to R^n$, $h: R^n \to R^{\ell}$, $k: R^n \times R^m \to R^{n_1}$, and $\Omega \subset R^m$; again we take control functions u in

$$U_{ad} := \{u \in L^{\infty}(T;R^m) : u(t) \in \Omega \text{ a.e.}\}.$$

This problem will turn out to be a special case of Problem V.1.1; note, however, that, compared to the periodic Problem V.1.2 we have added an "isoperimetric" constraint. Such constraints are often imposed in optimal periodic control problems. One has to take some care in stating the constraint qualifications.

The corresponding *optimal steady state problem* has the following form.

<u>Problem 1.2</u> Minimize $g(x,u)$ over $(x,u) \in R^n \times R^m$

$$\begin{array}{ll} \text{s.t.} & 0 = f(\bar{x},u) \\ & h(x) \in R^{\ell}_{-} \\ & 0 = k(x,u) \\ & u \in \Omega \ , \end{array}$$

here f, g, h, k, and Ω are as in Problem 1.1 and $\bar{x} \in C(-r,0;R^n)$ is the constant function $\bar{x}(s) \equiv x$.

One observation on the relation between these two problems is immediate: If $(x,u) \in R^n \times R^m$ satisfies the constraints of Problem 1.2, then the constant functions $(\bar{x},\bar{u}) \in C(-r,\tau,R^n) \times L^{\infty}(0,\tau,R^m)$, defined by $\bar{x}(t) \equiv x$, $\bar{u}(t) \equiv u$, satisfy the constraints of Problem 1.1.

Similarly as we distinguished between strong and weak local optimal solutions for the dynamic optimization problems considered in Chapters IV and V, one might distinguish *strong* and *weak local optimal solutions* of Problem 1.2: strong local solutions (x^0,u^0) have to be compared with all (x,u) satisfying the constraints of Problem 1.2 with $|x-x^0| < \varepsilon$ for some $\varepsilon > 0$, while for weak local solutions the additional constraint $|u-u^0| < \varepsilon$ is imposed.

We will not consider strong local solutions of Problem 1.2. Instead we restrict our attention to weak local solutions of Problem 1.2, which we call simply *local optimal solutions*.

Definition 1.3 A local solution $(x^0,u^0) \in R^n \times R^m$ of Problem 1.2 is called *locally proper*, if for all $\varepsilon > 0$ there exists a pair $(x,u) \in C(-r,\tau;R^n) \times L^\infty(0,\tau;R^m)$ with $\sup_{t\in T}|x^0-x(t)| < \varepsilon$ and $\sup_{t\in T}|u^0-u(t)| < \varepsilon$, satisfying the constraints of Problem 1.1 and

$$1/\tau \int_0^\tau g(x(t),u(t))dt < g(x^0,u^0).$$

If Ω is compact, we can consider the *Relaxed Problem 1.1*, by inserting relaxed controls $v \in S$ instead of u in the functions g, h and k. Similarly, one can consider the *Relaxed Problem 1.2* by inserting instead of $u \in \Omega$ a Radon probability measure v on Ω, $v \in rpm(\Omega)$. Accordingly, one has the following analogue of Definition 1.3.

Definition 1.4 A local solution $(x^0,v^0) \in R^n \times rpm(\Omega)$ of the *Relaxed Problem 1.2* is called *locally proper*, if for all $\varepsilon > 0$ there exists a pair $(x,v) \in C(-r,\tau;R^n) \times S$ with $\sup_{t\in T}|x^0 - x(t)| < \varepsilon$, satisfying the constraints of the Relaxed Problem 1.1 and

$$1/\tau \int_0^\tau g(x(t),v(t))dt < g(x^0,v^0).$$

This chapter analyzes tests for local properness.

2. Analysis of First Order Conditions

We will show that the first order necessary optimality conditions of Section V.2 do not yield a test for local properness since (modulo a constraint qualification) every local optimal solution (x^0,u^0) of the steady state Problem 1.2 satisfies these conditions. A similar result is valid for the first order necessary optimality conditions of Section VI.3, which are satisfied by every local solution of the steady state Relaxed Problem 1.2. Note, however, that local solutions of Problem 1.2, in general, do *not* satisfy the first order necessary conditions for the *Relaxed* Problem 1.2.

The attainability cone for Problem 1.1 has the form

$$A := \{(x_\tau-\varphi,z) \in C(-r,0;R^n)\times R^{n_1} : \varphi \in C(-r,0;R^n) \tag{2.1}$$

and there exists $u \in L^\infty(0,\tau;R^m)$ with

$u^o(t) + \alpha u(t) \in \Omega$ a.e. for some $\alpha > 0$, such that $x_o = \varphi$

$\dot{x}(t) = \mathcal{D}_1 f(x_t^o, u^o(t))x_t + \mathcal{D}_2 f(x_t^o, u^o(t))u(t)$ a.a. $t \in T$

$z = \int_0^\tau [k_x(x^o(t),u^o(t))x(t) + k_u(x^o(t),u^o(t))u(t)]dt \}$.

The cone A contains a subspace of finite codimension in $C(-r,0;R^n)\times R^{n_1}$. We formulate the following constraint qualification.

There exist $\tilde{u} \in \mathcal{U}_{ad}(u^o)$ and a solution $\tilde{x}$ of (2.2)

$\tilde{x}_o = \tilde{x}_\tau$, $\dot{\tilde{x}}(t) = \mathcal{D}_1 f(x_t^o, u^o(t))\tilde{x}_t + \mathcal{D}_2 f(x_t^o, u^o(t))\tilde{u}(t)$ a.a. $t \in T$

with

$0 = \int_0^\tau [k_x(x^o(t),u^o(t))x(t) + k_u(x^o(t),u^o(t))\tilde{u}(t)]dt$,

$h_x(x^o(t))x(t) \in \operatorname{int} R_-^\ell$ for all $t \in T$.

<u>Theorem 2.1</u> Let $(x^o,u^o) \in R^n\times R^m$ be such that the constant functions $(\bar{x}^o,\bar{u}^o) \in C(-r,\tau;R^n) \times L^\infty(0,\tau;R^m)$ are a weak local minimum of Problem 1.1 and assume

The maps f, g, h, and k are continuously Fréchet differentiable and bounded together with their derivatives for bounded arguments; the set Ω is closed and convex. (2.3)

Then there exist $\lambda_o \geq 0$, $y_1 \in R^{n_1}$, non-negative, regular Borel measures μ_i on T supported on the sets $\{t:h^i(x^o(t)) = 0\}$, $i = 1,\dots,\ell$, and a τ-periodic solution y of the adjoint equation

$$\frac{d}{ds}\{y(s) - \int_s^{\tau+r} \mathcal{D}_1 f(\bar{x}^o,u^o)^T y^t dt + \int_s^\tau [d\mu(t)]h_x(x^o)\} = -\lambda_o g_x(x^o,u^o) - y_1^T k_x(x^o,u^o) \quad \text{a.a. } s \leq \tau \tag{2.4}$$

such that $(\lambda_o, y^\tau, y_1, \mu) = \lambda \neq 0$ in $R\times NBV(0,r;R^n) \times R^{n_1}\times C(T;R^\ell)^*$ and

$$[\lambda_o g_u(x^o,u^o) + y(s)^T f_u(\bar{x}^o,u^o) + y_1^T k_u(x^o,u^o)][u-u^o] \geq 0 \tag{2.5}$$

for all $u \in \Omega$ and a.a. $s \in T$.

If (2.2) holds and A defined in (2.1) satisfies $A = C(-r,0;R^n)\times R^{n_1}$ then $\lambda_o \neq 0$.

<u>Proof:</u> Using a standard device in optimal control theory, augment the

state space to R^{n+n_1} by

$$\dot{x}^{n+j} = k_j(x(t),u(t)), \quad t \in T \tag{2.6}$$

with boundary condition

$$x^{n+j}(0) = x^{n+j}(\tau) = 0, \quad j = 1,\dots,n_1.$$

Then Theorem V.2.4 yields the result, if one notes that A is the attainable cone of the augmented system, and A has finite codimension in $C(-r,0;R^n)\times R^{n_1}$. Note that the adjoint equation is also obtained for $\tau < r$. □

First order necessary optimality conditions for Problem 1.2 have the following form.

Theorem 2.2 Suppose that $(x^0,u^0) \in R^n\times R^m$ is a local minimum of Problem 1.2 and condition (2.3) holds.

Then there exist $(\lambda_0,y,y_1,z) \in R_+\times R^n\times R^{n_1}\times R_-^{\ell}$, not all vanishing, with

$$\lambda_0 g_x(x^0,u^0) - y^T\mathcal{D}_1 f(\overline{x}^0,u^0) - y_1^T k_x(x^0,u^0) - z^T h_x(x^0) = 0 \tag{2.7}$$

$$[\lambda_0 g_u(x^0,u^0) - y^T f_u(\overline{x}^0,u^0) - y_1^T k_u(x^0,u^0)][u-u^0] \geq 0 \tag{2.8}$$

for all $u \in \Omega$; if

$$R^n\times R^{n_1} = \{(\mathcal{D}_1 f(\overline{x}^0,u^0)\overline{x} + f_u(\overline{x}^0,u^0)u, k_x(x^0,u^0)x + k_u(x^0,u^0)u) : x \in R^n,\ u \in \Omega(u^0)\} \tag{2.9}$$

and

$$R^{\ell} = \{h_x(x^0)x : \text{there exists } u \in \Omega(u^0) \text{ with } 0 = \mathcal{D}_1 f(\overline{x}^0,u^0)\overline{x} + f_u(\overline{x}^0.u^0)u \text{ and } 0 = k_x(x^0,u^0)x + k_u(x^0,u^0)u\} - R_-^{\ell}(h(x^0))\}. \tag{2.10}$$

then $\lambda_0 \neq 0$.

Proof: This follows from Theorem II.1.11. □

Remark 2.3 If Ω is compact and the constraints determine x corresponding to u uniquely, then Problem 1.2 has a solution.

Corollary 2.4 Suppose that $(x^0,u^0) \in R^n\times R^m$ is a local solution of Problem 1.2 and condition (2.3) holds.

Then $(\bar{x}^0,\bar{u}^0) \in C(-r,\tau;R^n) \times L^\infty(0,\tau;R^m)$ satisfy the first order conditions (2.4) and (2.5) for a weak local minimum of Problem 1.1; and for $(\lambda_0,y,y_1,z) \in R_+ \times R^n \times R^{n_1} \times R^\ell$ with (2.7) and (2.8) the solution $y(\cdot)$ of the adjoint equation satisfies

$$y^\tau(s) = -y, \quad 0 \le s < r \quad \text{and} \quad \mu = Z_0 z \tag{2.11}$$

where Z_0 is the embedding of R^ℓ into $C(T;R^\ell)^*$ given by

$$(Z_0 z)x = -\int_T z^T x(t)dt, \quad x \in C(T;R^\ell). \tag{2.12}$$

<u>Proof:</u> Follows from Theorems 2.1 and 2.2. □

Next we study the relation between the constraint qualifications for the periodic Problem 1.1 and the steady state Problem 1.2. We confine our attention to the case without state and control constraints and without isoperimetric constraint.

The linearized system equation has the form

$$\dot{x}(t) = Lx_t + B_0 u(t) \quad \text{a.a.} \quad t \in [0,\tau], \tag{2.13}$$

where $L := \mathcal{D}_1 f(\bar{x}^0,u^0)$ and $B_0 := \mathcal{D}_2 f(\bar{x}^0,u^0)$.

The condition for A in Theorem 2.1 specializes to

$$A := \{x_\tau - \varphi : \varphi \in C(-r,0;R^n) \text{ and there exists } u \in L^\infty(0,\tau;R^m) \text{ such that } x_0 = \varphi \text{ and } x \text{ solves (2.13)}\} = C(-r,0;R^n) \tag{2.14}$$

and (2.9) means here

$$R^n = \{L\bar{x} + B_0 u : x \in R^n,\ u \in R^m\}. \tag{2.15}$$

We want to make use of some notions from the theory of autonomous linear retarded systems in the state space $M^2 = R^n \times L^2(-r,0;R^n)$ (see e.g. Manitius [1981]).

Equation (2.13) induces a strongly continuous semigroup $S(t)$, $t \ge 0$, of operators on M^2. For $\varphi \in M^2$, let φ^0, φ^1 denote its R^n and $L^2(-r,0;R^n)$ components, respectively. Let $x(t)$ be a solution of (2.13) corresponding to some initial condition $x(0) = \varphi^0$, $x(\theta) = \varphi^1(\theta)$, $\theta \in [-r,0)$, where $\varphi \in M^2$, and to some control $u \in L^2(0,\tau;R^m)$. Then $z(t) = (x(t),x_t) \in M^2$ is the mild solution of the abstract differential equation

$$z(0) = \varphi, \quad \dot{z}(t) = Az(t) + Bu(t), \quad t \ge 0,$$

where $A: \mathcal{D}(A) \subset M^2 \to M^2$ is the infinitesimal generator of $S(t)$, $t \geq 0$, and $B: R^m \to M^2$ is the bounded linear operator $Bu := (B_0 u, 0)$; here the domain $\mathcal{D}(A)$ of A is the image of $W_2^{(1)}(-r,0;R^n)$ under the natural embedding and $A\varphi = (L\varphi, \dot{\varphi})$. Let $\Delta(\lambda)$ be the characteristic matrix

$$\Delta(\lambda) = \lambda I - L(e^{\lambda \cdot}).$$

We recall that the spectrum of A is $\sigma(A) := \{\lambda \in \mathbb{C} \mid \det \Delta(\lambda) = 0\}$ and $\rho(A) := \mathbb{C} \setminus \sigma(A)$ is the resolvent set of A. For $\lambda \in \sigma(A)$ let M_λ denote the generalized eigenspace of A corresponding to λ, that is $M_\lambda = \bigcup_{k \in N} \ker(\lambda I - A)^k$.

Definition 2.5 The generalized eigenspace M_λ is called controllable if the canonical projection of (2.12) on M_λ is completely controllable.

We obtain the following interpretation of (2.15).

Proposition 2.6 Condition (2.15) holds iff $\lambda = 0$ is in the resolvent set $\rho(A)$ or the generalized eigenspace M_0 is controllable.

Proof: Remember that M_λ is controllable iff

$$n = \text{rank}\, [\Delta(\lambda), B_0] = \text{rank}\, [\lambda I - L(e^{\lambda \cdot}), B_0].$$

For $\lambda = 0$, this means

$$\{L\bar{x} \mid \bar{x} \in R^n\} + \text{Im}\, B_0 = R^n, \quad \text{i.e. (2.15)}. \qquad \square$$

For systems governed by ordinary differential equations, the normality conditions which are obtained by a specialization of (2.14) and (2.15), respectively, are equivalent. However the proof of this result given in Bernstein/Gilbert [1980,Theorem 4.3] breaks down for functional differential systems, as can be seen from the following discussion.

One can easily show (cp. Manitius [1981,p.531]) that the condition $\text{rank}[\Delta(\lambda), B_0] = n$ is equivalent to

$$\text{Im}(\lambda I - A) + \text{Im}\, B = M^2.$$

Hence (2.15) is equivalent to

$$\text{Im}\, A + \text{Im}\, B = M^2. \tag{2.16}$$

Furthermore (cp. e.g. Pazy [1983,p.5])

$$[S(\tau) - Id_M]z = A \int_0^\tau S(\sigma) z d\,, \quad z \in M^2. \tag{2.17}$$

Hence $\operatorname{Im}(S(\tau)-Id_M) \subset \operatorname{Im} A$.

Let the subspace attainable from zero be defined by

$$A^o := \{(x(\tau),x_\tau) : \text{there exists } u \in L^\infty(0,\tau;R^m) \text{ such that}$$
$$x_0 = 0,\ \dot{x}(t) = Lx_t + B_0u(t), \text{ a.a. } t \in [0,\tau]\}.$$

Then (2.14) means

$$A = \operatorname{Im}[S(\tau)-Id_C] + A^o = C(-r,0;R^n) \subset M^2. \tag{2.18}$$

Hence (2.14) implies by (2.17).

$$\operatorname{Im} A + A^o \supset C(-r,0;R^n). \tag{2.19}$$

But $A^o \subset W_2^{(1)}(-r,0;R^n) \subset M^2$. Hence $A^o \cap \operatorname{Im} B = 0$. Thus there is no way to conclude from (2.19) that (2.16) holds. Contrarily, for ordinary differential systems, $A = \operatorname{Im}[B,AB,\dots,A^{n-1}B]$. Hence (2.19) is equivalent to $\operatorname{Im} A + \operatorname{Im} B = R^n$, i.e. (2.15).

Conversely observe that by definition of the integral

$$\int_0^\tau S(\sigma)z\,d\sigma \in \text{closure} \bigcup_{\sigma\in(0,\tau)} \operatorname{Im} S(\sigma).$$

For general delay systems, this is a proper subset of $\mathcal{D}(A)$. Hence (2.17) does not imply $\operatorname{Im}[S(\tau)-Id_M] = \operatorname{Im} A$. For ordinary differential equations, however, (2.17) implies that $\operatorname{Im}[S(\bar\tau)-Id] = \operatorname{Im}[e^{A\bar\tau}-I] = \operatorname{Im} A$ for all but a finite number of values of $\bar\tau$ in $[0,\tau]$. Then we conclude from (2.15) that

$$\begin{aligned} R^n &= \operatorname{Im} A + \operatorname{Im} B \\ &= \operatorname{Im} A + \operatorname{Im}[B,AB,\dots,A^{n-1}B] \\ &= \operatorname{Im}[e^{A\bar\tau}-I] + A^o \end{aligned}$$

i.e. (2.14) follows for all but a finite set of values of $\bar\tau$ in $[0,\tau]$.

In Section 3, below, we will need more information than that furnished by Corollary 2.4: *Every* Lagrange multiplier from Problem 1.1 must be obtainable in the form (2.11). In presence of the state constraint, this can - under additional assumptions - be achieved on the basis of Theorem II.1.11. In this section, we exclude state constraints; in Section IX.3 we will discuss ordinary differential equations subject to state constraints.

Proposition 2.7 Consider Problem 1.1 without state constraint (i.e. $h \equiv 0$). Suppose that for $(x^0,u^o) \in R^n \times R^m$ the constant functions $(\bar{x}^0,\bar{u}^o) \in C(-r,\tau;R^n) \times L^\infty(0,\tau;R^m)$ are a weak local minimum, condition

(2.3) is satisfied and the homogeneous linearized system equation (2.13) has only the trivial τ-periodic solution. Then

(i) every τ-periodic solution y of the adjoint equation

$$\frac{d}{ds}\, y(s) + \mathcal{D}_1 f(\bar{x}^o,u^o)^T y(s) = -\lambda_o g_x(x^o,u^o) - k_x(x^o,u^o)^T y_1, \quad s\in R \tag{2.20}$$

is constant;

(ii) every Lagrange multiplier $\lambda = (\lambda_o, y^*, y_1) \in R_+ \times C(-r,0;R^n)^* \times R^{n_1}$ (satisfying (2.4) and (2.5)) for Problem 1.1 is of the form (2.11).

Proof: The assumptions imply that $z = jk\omega$, $k \in Z$, $\omega := 2\pi/\tau$ is not an eigenvalue of (2.13), i.e. the characteristic function

$$\Delta(z) := zI - \mathcal{D}_1 f(\bar{x}^o,u^o)(e^{z\cdot}) \tag{2.21}$$

satisfies in particular

$$\begin{aligned} n &= \text{rank}\ \Delta(0) \\ &= \text{rank}\ \mathcal{D}_1 f(\bar{x}^o,u^o) \\ &= \text{rank} \int_{-r}^{0} [d\eta(s)]. \end{aligned} \tag{2.22}$$

Thus the equation

$$y^T \mathcal{D}_1 f(\bar{x}^o,u^o) = \lambda_o g_x(x^o,u^o) - y_1^T k_x(x^o,u^o) \tag{2.23}$$

has a unique solution $y \in R^n$.

This means that the adjoint equation (2.20) has the constant solution $\bar{y}(s) \equiv y$. On the other hand, the assumptions imply that (2.20) has a unique τ-periodic solution. This proves (i), and (ii) follows by Corollary 2.4.

□

The proposition above shows that, in order to get a one-one correspondence between Lagrange multipliers of Problems 1.1 and 1.2, one only has to require a constraint qualification for the dynamic constraint.

Results completely analogous to Theorems 2.1, 2.2 and Proposition 2.7 hold for the *Relaxed Problems 1.1 and 1.2*.

We content ourselves with stating for later use the result analogous to Proposition 2.7.

Proposition 2.8 Suppose that for $(x^o,v^o) \in R^n \times rpm(\Omega)$ the constant functions $(\bar{x}^o,\bar{v}^o) \in C(-r,\tau;R^n) \times S$ are a local minimum of the Relaxed Problem 1.1 (with $h = k \equiv 0$). Assume

The maps f and g are continuously Fréchet differentiable (2.24)
with respect to their first arguments; for all $u \in \Omega$ and all bounded x and φ, $g(x,u)$, $g_x(x,u)$, $f(\varphi,u)$, and $\mathcal{D}_1 f(\varphi,u)$ are bounded; and the set Ω is compact.

The linearized equation (2.25)

$$\dot{x}(t) = \mathcal{D}_1 f(\bar{x}^0, v^0) x_t \quad \text{a.a.} \quad t \in R$$

has only the trivial τ-periodic solution.

Then

(i) every τ-periodic solution of the adjoint equation

$$\frac{d}{ds} y(s) + \mathcal{D}_1 f(\bar{x}^0, v^0)^T y^s = -\lambda_0 g_x(x^0, v^0) \quad \text{a.a.} \quad s \in R \tag{2.26}$$

is constant.

(ii) Hence every Lagrange multiplier $(\lambda_0, y^*) \in R_+ \times Y^*$ (satisfying conditions (VI.3.8) and VI.3.9)) for the Relaxed Problem 1.1 is of the form (2.11) for Lagrange multipliers (λ_0, y, y_1) of the Relaxed Problem 1.2.

Corollary 2.4 shows that every local optimal solution of the steady state Problem 1.2 satisfies the first order necessary optimality conditions for the dynamic Problem 1.1. This result can better be understood in the light of Proposition II.1.14. The rest of this section is devoted to this discussion.

In the Banach space reformulation as Problem V.1.12, Problem 1.1 (where for simplicity $h = k \equiv 0$ and we assume that all maps are globally defined) is equivalent to

Problem 2.9 Minimize $G(S(u,\varphi),u)$

over all $(u,\varphi) \in L^\infty(0,\tau;R^m) \times C(-r,0;R^n) =: X$

satisfying

$$S(u,\varphi)_\tau - \varphi = 0 \tag{2.27}$$

$$(u,\varphi) \in \mathcal{U}_{ad} \times C(-r,0;R^n) =: C; \tag{2.28}$$

here S is the solution operator of

$$x_o = \varphi, \quad \dot{x}(t) = f(x_t, u(t)) \quad \text{a.a.} \quad t \in [0,\tau]. \tag{2.29}$$

Let

$$\mathfrak{X} := \{(\bar{u},\bar{x}) \in L^\infty(0,\tau;R^m) \times C(-r,0;R^n) : u \in R^m,\ x \in R^n\},$$

where $\overline{u}$ and $\overline{x}$ denote the constant functions $\overline{u} \equiv u$, $\overline{x} \equiv x$. We identify $\widetilde{X}$ with $R^m \times R^n$. Then Problem 1.2 can, in a rather hypertrophic manner, be rewritten as

<u>Problem 2.10</u> Minimize $G(S(u,\varphi),u)$

over all $(u,\varphi) \in \widetilde{X}$ satisfying

$$S(u,\varphi)_\tau - \varphi = 0 \tag{2.30}$$

$$(u,\varphi) \in \widetilde{X} \cap (\mathcal{U}_{ad} \times C(-r,0;R^n)) := \widetilde{C}. \tag{2.31}$$

In fact, let $(\overline{u},\overline{x}) \in \widetilde{C}$. Then (2.23) implies for $\hat{x} := S(\overline{u},\overline{x})$

$$\hat{x}(t) = \hat{x}(\tau-r) + \int_{\tau-r}^{t} f(\hat{x}_s,u)ds$$
$$= x \qquad , \quad t \in [\tau-r,\tau]$$

whence

$$\hat{x}(\tau) = x = x + \int_{\tau-r}^{\tau} f(\overline{x},u)ds.$$

Thus

$$f(\overline{x},u) = 0$$

and hence $\hat{x}(t) = x$ for all $t \in [0,\tau]$.

Hence every pair (x,u) satisfying the constraints of Problem 2.10 satisfies the constraints of Problem 1.2. The converse is obvious.

Define a linear map $P = (P_u,P_\varphi) : X \to \widetilde{X}$ by

$$P(u,\varphi) = (1/\tau \int_0^\tau u(t)dt,\ 1/r \int_{-r}^0 \varphi(t)dt)). \tag{2.32}$$

Observe $P C \subset \widetilde{C}$, since Ω is convex. Furthermore, for constant u,φ one has $P(u,\varphi) = (u,\varphi)$.

Now let $(x^0,u^0) \in R^n \times R^m$ be a local optimal solution of Problem 1.2, and hence of Problem 2.10. Define $\varphi^0 := \overline{x}^0$. Then

$$(\overline{u}^0,\varphi^0) \in \widetilde{C} \text{ and } S(\overline{u}^0,\varphi^0) = \overline{x}^0.$$

Recall that the Lagrangean for Problem 2.10 is given by (V.2.3):

$$L(u,\varphi,\lambda) = \lambda_0 G(S(u,\varphi),u) - y^* P(S(u,\varphi),\varphi)$$
$$= \lambda_0 G(S(u,\varphi),u) - y^*[S(u,\varphi)_\tau - \varphi]$$

where $\lambda = (\lambda_0,y^*) \in R_+ \times NBV(-r,0;R^n)$.

By Theorem 2.2 there are Lagrange multipliers $(\lambda_0, y) \in R_+ \times R^n$ for $(x^0, u^0) \in R^n \times R^m$ as a solution of Problem 1.2. Define $\lambda = (\lambda_0, y^*) := (\lambda_0, F^* y) \in R_+ \times NBV(-r,0;R^n)$ (cp.(2.11)).

Thus y is a constant solution of (V.2.22).

We compute, recalling (V.2.12),

$$\begin{aligned}
&\mathcal{D}_1 L(u^0,\varphi^0,\lambda)u \\
&= \lambda_0 \mathcal{D}_1 G(S(u^0,\varphi^0),u^0)\mathcal{D}_1 S(u^0,\varphi^0)u + \lambda_0 \mathcal{D}_2 G(S(u^0,\varphi^0),u^0)u \\
&\qquad - y^*[\mathcal{D}_1 S(u^0,\varphi^0)u] \\
&= \lambda_0/\tau \int_0^\tau g_x(x^0,u^0) \int_0^t \Phi(t-s)X_0 \mathcal{D}_2 f(\bar{x}^0,u^0)u(s)\,ds\,dt \\
&+ \lambda_0/\tau \int_0^\tau g_u(x^0,u^0)u(t)\,dt - (F^*Y_0 y)\Big(\int_0^\tau \Phi(\tau-s)X_0 \mathcal{D}_2 f(\bar{x}^0,u^0)u(s)\,ds\Big) \\
&= \lambda_0/\tau \int_0^\tau g_u(x^0,u^0)u(t)\,dt + \int_0^\tau y^T \mathcal{D}_2 f(\bar{x}^0,u^0)u(t)\,dt \\
&= \lambda_0/\tau \int_0^\tau g_u(x^0,u^0)(P_u u)\,dt + \int_0^\tau y^T \mathcal{D}_2 f(\bar{x}^0,u^0)(P_u u)\,dt \\
&= \mathcal{D}_1 L(u^0,\varphi^0,\lambda)P_u u.
\end{aligned}$$

Similarly,

$$\begin{aligned}
&\mathcal{D}_2 L(u^0,\varphi^0,\lambda)\varphi \\
&= \mathcal{D}_2 L(u^0,\varphi^0,\lambda)(P_\varphi \varphi) = 0.
\end{aligned}$$

Thus the assumptions of Proposition II.1.14 are verified.

This discussion shows that first order necessary optimality conditions for Problems 1.1 and 1.2 coincide, since the steady state Problem 1.2 can be obtained by "projection" of the periodic Problem 1.1.

3. The Π-Test

In this section we use the second order necessary conditions for a weak local minimum derived in Section V.3 in order to get tests for local properness. Throughout, we consider Problems 1.1 and 1.2 *without* state constraints.

Under the assumptions of Theorem 2.2, let Λ be the following set of Lagrange multipliers for the steady state Problem 1.2:

$$\Lambda := \{0 \neq \lambda = (\lambda_0, y, y_1) \in R_+ \times R^n \times R^{n_1} : \quad (2.7) \text{ and } (2.8) \text{ hold}, |\lambda| \leq 1\}. \tag{3.1}$$

Define $L := \mathcal{D}_1 f(\bar{x}^0, u^0)$ and

$$B := \mathcal{D}_2 f(\bar{x}^0, u^0) \tag{3.2}$$

and the characteristic function

$$\Delta(z) = zI - L(e^{-z\cdot}), \quad z \in C \tag{3.3}$$

where I is the $n\times n$ unit matrix.
Let the Pontryagin function H be

$$H(\varphi, u, \lambda) = \lambda_0 g(\varphi(0), \omega) + y^T f(\varphi, u) + y_1^T k(\varphi(0), u). \tag{3.4}$$

<u>Theorem 3.1</u> Suppose that $(x^0, u^0) \in R^n \times R^m$ is a local solution of Problem 1.2 with $h \equiv 0$ and assume, in addition to (2.3)

the maps f, g, and k are twice continuously Frêchet differentiable and the first and second derivatives are bounded for bounded arguments; (3.5)

$$\operatorname{rank} \Delta(jk\omega) = n \quad \text{for all} \quad k \in Z \tag{3.6}$$

where $\omega := 2\pi/\tau$.
Then (x^0, u^0) is locally proper if there exists $(x,u) \in C(-r,\tau;R^n) \times L^\infty(0,\tau;R^m)$ with

$$\int_0^\tau [g_x(x^0,u^0)x(t) + g_u(x^0,u^0)u(t)]dt \leq 0 \tag{3.7}$$

$$x_0 = x_\tau, \quad \dot{x}(t) = Lx_t + Bu(t)), \quad \text{a.a. } t \in [0,\tau] \tag{3.8}$$

$$u \in \mathcal{U}_{ad}(u^0)$$

$$\max_{\lambda\in\Lambda} \int_0^\tau [\mathcal{D}_1\mathcal{D}_1 H(\bar{x}^0,u^0,\lambda)(x_t,x_t) + 2\mathcal{D}_1\mathcal{D}_2 H(\bar{x}^0,u^0,\lambda)(x_t,u(t)) + \mathcal{D}_2\mathcal{D}_2 H(\bar{x}^0,u^0,\lambda)(u(t),u(t))]dt < 0. \tag{3.9}$$

<u>Proof:</u> This is a simple consequence of Theorem V.3.8 and Proposition 2.7. Note that Λ is closed and bounded, hence compact and the maximum exists. □

<u>Remark 3.2</u> Condition (3.9) involves only Lagrange multipliers for the finite dimensional Problem 1.2. In spite of this it furnishes a

(negative) test for optimality in the infinite dimensional Problem 1.1.

Due to condition (3.6) the periodic solution x in (3.8) is uniquely determined by $u(\cdot)$. In fact one can compute the Fourier coefficients of x using those of u. This is the basis for the so-called Π-Test, which restricts the set of test functions u. This simplified version can be used effectively to analyse local properness in dependence of the frequency ω (i.e. the period length τ).

Let u be an element of L^2_τ, i.e. an equivalence class of τ-periodic functions with

$$|u| := 1/\tau \int_0^\tau u(t)^2 dt < \infty$$

(cp. for this and the following e.g. Butzer/Nessel [1971]).
Then u has the expansion

$$u(t) = \sum_{k=-\infty}^{\infty} x\hat{}(k) e^{jk\omega t} \tag{3.10}$$

where $j := \sqrt{-1}$ and

$$x\hat{}(k) := 1/\tau \int_0^\tau u(t) e^{-jk\omega t} dt = 1/2\pi \int_{-\pi}^{\pi} u(\tfrac{t}{\omega}) e^{-jkt} dt. \tag{3.11}$$

and the corresponding solution x of (3.8) has the expansion

$$x(t) = \sum_{k=-\infty}^{\infty} x\hat{}(k) e^{jk\omega t} \tag{3.12}$$

where

$$x\hat{}(k) = \Delta^{-1}(jk\omega) B u\hat{}(k) = T(jk\omega) u\hat{}(k), \tag{3.13}$$

defining the transfer function $T(\cdot)$ by

$$T(z) = \Delta^{-1}(z) B, \quad z \in C. \tag{3.14}$$

Since the trajectory x is absolutely continuous, hence of bounded variation and continuous, convergence in (3.12) is uniform, Edwards [1967,Remark 2,p.151].

For u satisfying (3.8) one can plug the series expansions of x and u into (3.9) in order to get a test involving only the Fourier coefficients $u\hat{}(k)$ of u, the transfer function $T(jk\omega)$ and H.

We will state this only for functions u with $u\hat{}(k) = 0$, $k \neq 0, \pm 1$. This is by far the most important case. The requirement (3.8) can e.g. be satisfied by choosing the constant ("zero frequency") term of u as leading into the interior of Ω and then choosing the next Fourier coefficient such that $u^o + u$ remains in Ω.

Abbreviate for $\omega \in R_+$

$$P(\omega,\lambda) := \mathcal{D}_1\mathcal{D}_1 H(\bar{x}^o,u^o,\lambda)(e^{j\omega\cdot}I, e^{-j\omega\cdot}I) \tag{3.15}$$

$$Q(\omega,\lambda) := \mathcal{D}_2\mathcal{D}_1 H(\bar{x}^o,u^o,\lambda)(e^{j\omega\cdot}I)$$

$$R(\lambda) := \mathcal{D}_2\mathcal{D}_2 H(\bar{x}^o,u^o,\lambda).$$

We identify $P(\omega,\lambda)$, $Q(\omega,\lambda)$ and $R(\lambda)$ with elements in $C^{n\times n}$, $C^{n\times m}$ and $R^{m\times m}$, respectively.

Observe that $R(\lambda)$ is symmetric, $P(\omega,\lambda)$ is Hermitian, and $Q(-\omega,\lambda) = \mathcal{D}_1\mathcal{D}_2 H(\bar{x}^o,u^o,\lambda_o,y,y_1)(e^{-j\omega\cdot})$.

Define $\Pi(\omega,\lambda) \in C^{m\times m}$ by

$$\Pi(\omega,\lambda) := T(-j\omega)^T P(\omega,\lambda) T(j\omega) \tag{3.16}$$

$$+ Q(-\omega,\lambda)^T T(j\omega) + T(-j\omega)^T Q(\omega,\lambda) + R(\lambda).$$

<u>Theorem 3.3</u> (Π-Test) Suppose that $(x^o,u^o) \in R^n \times R^n$ is a local solution of Problem 1.2 with $h \equiv 0$ and assume that conditions (2.3), (3.5), and (3.6) hold.

Then (x^o,u^o) is locally proper, if there exist $\nu_o,\nu_1 \in R^m$ with

$$[g_x(x^o,u^o)T(0) + g_u(x^o,u^o)]\nu_o \tag{3.17}$$

$$+ [g_x(x^o,u^o)T(j\omega) + g_u(x^o,u^o)]\nu_1 \le 0$$

$$u^o + \nu_o + \operatorname{Re} \nu_1 e^{j\omega t} \in \Omega \quad \text{f.a.} \quad t \in [0,\tau]$$

$$\max_{\lambda\in\Lambda} [\nu_o^T \Pi(o,\lambda)\nu_o + 2\nu_1^T \Pi(\omega,\lambda)\nu_1] < 0. \tag{3.18}$$

where Λ defined by (3.1).

<u>Proof:</u> The function

$$u(t) := \nu_o + \operatorname{Re} \nu_1 e^{j\omega t}, \quad t \in [0,\tau]$$

satisfies $u \in \mathcal{U}_{ad}(u^o)$ and has Fourier coefficients

$$u\hat{}(0) = \nu_o$$

$$u\hat{}(\pm 1) = \nu_1/2, \; u\hat{}(k) = 0, \quad k \neq 0, \; \pm 1.$$

The corresponding solution x of (3.8) has, by (3.13), Fourier coefficients

$$x\hat{}(0) = T(0)\lambda^0$$

$$x\hat{}(1) = T(j\omega)\nu_1/2$$

$$x\hat{}(-1) = T(-j\omega)\nu_1/2$$

$$x\hat{}(k) = 0 \qquad , \quad k \neq 0, \ \pm 1.$$

We compute, using orthonormality of $\{e^{jk\cdot}, k \in Z\}$

$$\int_0^\tau \mathcal{D}_1\mathcal{D}_1 H(\overline{x}^0,u^0,\lambda)(x_t,x_t)dt$$

$$= \int_0^\tau \mathcal{D}_1\mathcal{D}_1 H(\overline{x}^0,u^0,\lambda)\Big(\sum_{k=-1}^{1} x\hat{}(k)e^{jk\omega(t+\cdot)}, \sum_{k=-1}^{1} x\hat{}(k)e^{jk\omega(t+\cdot)}\Big)dt$$

$$= \int_0^\tau \sum_{k,k'=-1}^{1} x\hat{}(k)^T e^{jk\omega t}\mathcal{D}_1\mathcal{D}_1 H(\overline{x}^0,u^0,\lambda)(e^{jk\omega\cdot}, \ e^{jk'\omega\cdot})x\hat{}(k')e^{jk'\omega t}dt$$

$$= x\hat{}(0)^T\mathcal{D}_1\mathcal{D}_1 H(\overline{x}^0,u^0,\lambda)(e^{0\cdot},e^{0\cdot})x\hat{}(0)$$

$$+ \ 2x\hat{}(-1)^T\mathcal{D}_1\mathcal{D}_1 H(\overline{x}^0,u^0,\lambda)(e^{-j\omega\cdot},e^{j\omega\cdot})x\hat{}(1)$$

$$= \nu_0^T \ T(0)^T\mathcal{D}_1\mathcal{D}_1 H(\overline{x}^0,u^0,\lambda)(e^{0\cdot},e^{0\cdot})T(0)\nu_0$$

$$+ \ 2\nu_1^T \ T(-j\omega)^T\mathcal{D}_1\mathcal{D}_1 H(\overline{x}^0,u^0,\lambda)(e^{-j\omega\cdot},e^{j\omega\cdot})T(j\omega)\nu_1 .$$

Similar expressions are obtained for the other terms in (3.9). Inserting the definition (3.16) of $\Pi(\omega)$ one obtains the assertion. □

Remark 3.4 If $u^0 \in \text{int}\ \Omega$, then ν_0 can be chosen as zero. If additionally, λ with $\lambda_0 = 1$ is unique, the familiar form of the Π-Test

$$\nu^T\Pi(\omega)\nu < 0 \tag{3.19}$$

is obtained (cp. Bittanti/Fronza/Guarbadassi [1973]).

The general form of Theorem 3.3 has been given by Bernstein [1984] for ordinary differential equations. See Remark IX.3.6 for more references in the case of ordinary differential equations.

Remark 3.5 Sincic/Bailey [1978] were the first to consider the Π-Test for delay equations. For problems without control constraints, they indicate the form of the Π-Test and give a formal proof (without taking into account constraint qualifications).

Their result is more general due to the consideration of state and control dependent delays. A rigorous proof in this case is still lacking due to differentiability problems.

Remark 3.6 Condition (3.19) holds for some $\nu \in R^m$ iff $\bar{\nu}^T \Pi(\omega)\nu < 0$ for some $\nu \in C^m$, since $\Pi(\omega)$ is Hermitian.

Though obtained by an apparent specialization of Theorem 3.1, the Π-Test is in many cases not weaker than Theorem 3.1. More precisely, the following result holds.

Proposition 3.7 Under the assumptions of Theorem 3.1, suppose that $\Omega = R^m$ and (x,u) are such that (3.7) and (3.9) hold. Then there exist $k \in Z$, $\nu \in C^m$ with

$$\max_{\lambda \in \Lambda} \bar{\nu}^T \Pi(k\omega,\lambda)\nu < 0$$

where $\omega = 2\pi/\tau$.

Proof: This follows using the Fourier expansions of x and u in (3.9): At least one summand of the obtained expression must be negative; however, each summand has the form

$$\bar{\nu}_k^T \Pi(k\omega,\lambda)\nu_k, \quad \nu_k \in C^m. \qquad \square$$

It is interesting to note that the Π-Test also implies the Legendre-Clebsch condition, obtained in Corollary IV.2.11 as a consequence of the global maximum principle.

Corollary 3.8 Let $(x^o,u^o) \in R^n \times R^n$ be a weak local minimum of Problem 1.1 for all $\tau > 0$ sufficiently small let $\Omega = R^m$ and assume that (3.5) holds and rank $\Delta(j\omega) = n$ for ω large enough. Then

$$\mathcal{D}_2\mathcal{D}_2 H(\bar{x}^o,u^o,\lambda) \geq 0$$

(i.e. this matrix is positive semi-definite).

Proof: Optimality implies that

$$\nu^T \Pi(\omega,\lambda)\nu \geq 0 \quad \text{for all} \quad \nu \in R^n \quad \text{and all} \quad \omega \in R \quad \text{large enough.}$$

(λ is unique here)

Note that only the first three summands in (3.16) contain $\Delta^{-1}(j\omega)$ as a factor, the other factor being bounded for $\omega \to \infty$.

But $\Delta^{-1}(z)$ is the Laplace transform of an integrable ($R^{n\times n}$-valued) function (see Kappel [1984,Theorem 5.7 and pp.10]). Hence the Lemma of Riemann-Lebesgue implies $\Delta^{-1}(z) \to 0$ for $|z| \to \infty$. For $\tau \to 0$, i.e. $\omega \to \infty$ we obtain the assertion.

□

Remark 3.9 The Π-Test is based on Fourier-series. For problems with control constraints, one might consider instead an orthonormal basis of L^2 consisting of piecewise constant functions, and use them as test functions (such a basis is given by Walsh-functions, cp. e.g. Tzafestas [1983]; see Rockey [1982] for an application to the solution of delay equations). Then, however, the corresponding τ-periodic solution x of the linearized system equation will have infinitely many non vanishing coordinates with respect to this basis.

It is helpful, to illustrate the Π-Test for the special case of a single, time-invariant delay; i.e. for the equation

$$\dot{x}(t) = f(x(t),x(t-r),u(t)) \tag{3.20}$$

where $f: R^n \times R^n \times R^m \to R^n$, and $r > 0$.
We assume that $f(x,y,u)$ is twice continuously differentiable in (x,y,u) and write

$$f_x, f_y, f_{xx} \quad \text{etc.}$$

for the partial derivatives.

Let $(x^0,u^0) \in R^n \times R^m$ be a steady state of (3.20), and suppose $\lambda_0 = 1$. Abbreviate

$$A_0 := f_x(x^0,x^0,u^0), \quad A_1 := f_y(x^0,x^0,u^0), \quad B := f_u(x^0,x^0,u^0). \tag{3.21}$$

In the following, all derivatives are evaluated at (x^0,x^0,u^0).

The Pontryagin function H is

$$H(x,y,u,\lambda) := g(x,u) + \lambda^T f(x,y,u) \tag{3.22}$$

for $(x,y,u,\lambda) \in R^n \times R^n \times R^m \times R^n$
(Note the change in the meaning of y, λ compared to (3.4)).
H_x, H_y, H_{xx} etc. denote partial derivatives of H evaluated at (x^0,x^0,u^0,λ).

By (3.15)

$$P(\omega) = H_{xx} + 2H_{xy}\exp(-j\omega r) + H_{yy} \tag{3.23}$$

$$Q(\omega) = H_{ux} + H_{uy}\exp(-j\omega r) \quad , \quad R = H_{uu}$$

and $T(\omega) = \Delta^{-1}(j\omega)f_u = \{j\omega I - A_o - A_1 \exp(-j\omega r)\}^{-1}B$.

Thus in this case one has

$$\begin{aligned}\Pi(\omega) = {} & B^T\{-j\omega I-A_o-A_1\exp(j\omega r)\}^{-1} P(\omega)\{j\omega I-A_o-A_1\exp(-j\omega r)\}^{-1}B \\ & + Q(-\omega)^T\{j\omega I-A_o-A_1\exp(-j\omega r)\}^{-1}B \\ & + B^T\{-j\omega I-A_o-A_1\exp(j\omega r)\}^{-1}Q(\omega) + R.\end{aligned} \tag{3.24}$$

In Section VIII.3 these formulae will be used for the analysis of an example.

4. The High Frequency Π-Test

The Π-Test requires (among other computations) that the transfer function $T = \Delta^{-1}f_u$ of the linearized system be explicitly known. For higher dimensional systems (see e.g. Watanabe/Onogi/Matsubara [1981] for a four-dimensional ordinary differential system arising in chemical engineering) this may demand considerable effort. The "high-frequency Π-Test" proven below does *not* require computation of the transfer function T. However, it is only applicable, if local properness occurs for all sufficiently high frequencies.

We consider throughout this section only the case without state or control constraint and without isoperimetric constraint (i.e. $\Omega = R^m$, $h = k \equiv 0$). Suppose that, in the situation considered in the last section, the following condition is satisfied:

$$\det \Delta(z) \text{ has no zero in the closed right half plane } \{z \in C : \operatorname{Re} z \geq 0\}. \tag{4.1}$$

Hence $\lambda_0 \neq 0$ in the optimality conditions, and the Lagrange multipliers with $\lambda_0 = 1$ are unique.

It follows from Kappel [1984, Proposition 5.7] that for $\operatorname{Re} z > -\delta$, $\delta > 0$, $\Delta^{-1}(z)$ is the Laplace transform of the fundamental solution $Y(t)$ of the equation in (3.7).

For $0 \neq \omega \in R$ we may write (cp. Kappel [1984, pp.10])

$$\Delta(j\omega) = j\omega(I-1/(j\omega) \int_{-r}^{0} \exp(j\omega\theta)d\eta(\theta))$$

and

$$\left|\frac{1}{j\omega} \int_{-r}^{0} \exp(j\omega\theta)d\eta(\theta)\right| \leq |L|/|\omega|.$$

Thus, for $|\omega| > |L|$, $\Delta^{-1}(j\omega)$ is given by

$$\Delta^{-1}(j\omega) = 1/(j\omega)\sum_{k=o}^{\infty}[(1/(j\omega)\int_{-r}^{o}\exp(j\omega\theta)d\eta(\theta))]^k \tag{4.2}$$

the series converging absolutely. Moreover, for any $\varepsilon > 0$ the series is uniformly converging for $|\omega| \geq |L| + \varepsilon$. Introduce for $0 \neq \omega \in R$ the $n\times n$-matrix $A(\omega)$ over C by

$$A(\omega) := \int_{-r}^{o}\exp(j\omega\theta)d\eta(\theta). \tag{4.3}$$

Then $\overline{A(\omega)} = A(-\omega)$, and

$$\Delta^{-1}(j\omega) = 1/(j\omega)\sum_{k=o}^{\infty}(j\omega)^{-k}A(\omega)^k \tag{4.4}$$

the series converging absolutely and uniformly for $|\omega| \geq |L| + \varepsilon$. Note that for all $\omega \in R_+$

$$|P(\omega)| \leq |\mathcal{D}_1\mathcal{D}_1 H(\overline{x}^0, u^o, \lambda)| < \infty$$

$$|Q(\omega)| \leq |\mathcal{D}_1\mathcal{D}_2 H(\overline{x}^0, u^o, \lambda)| < \infty.$$

Inserting (4.4) into (3.16), one obtains for $|\omega| > |L| + \varepsilon$

$$\begin{aligned}\Pi(\omega) &= B^T\sum_{i=o}^{\infty}(-j\omega)^{-i-1}A^T(-\omega)^i P(\omega)\sum_{\ell=o}^{\infty}(j\omega)^{-\ell-1}A(\omega)^{\ell}B \\ &\quad + Q^T(-\omega)\sum_{k=o}^{\infty}(j\omega)^{-k-1}A(\omega)^k B \\ &\quad + B^T\sum_{k=o}^{\infty}(j\omega)^{-k-1}A^T(-\omega)^k Q(\omega) + R \\ &= \sum_{k=o}^{\infty}(j\omega)^{-k-1}\{-\sum_{\substack{i+\ell=k\\ i,\ell>o}} B^T(-A^T(-\omega))^i[P(\omega)/(j\omega)]A(\omega)^{\ell}B \\ &\qquad + Q^T(-\omega)A(\omega)^k B - B^T(-A^T(-\omega))^k Q(\omega)\} + R.\end{aligned}$$

Define

$$R_o := R$$

$$R_k := [Q^T(-\omega)\ \ B^T]\begin{bmatrix} A(\omega) & 0 \\ -P(\omega)/(j\omega) & -A^T(-\omega)\end{bmatrix}^k\begin{bmatrix} B \\ -Q(\omega)\end{bmatrix} \tag{4.5}$$

By induction one finds

$$R_k(\omega) = [Q^T(-\omega)\ B^T]\begin{bmatrix} A(\omega)^k & 0 \\ -\sum\limits_{\substack{i+\ell=k-1\\ i,\ell\geq o}} (-A^T(-\omega))^i P(\omega)/(j\omega)A(\omega)^\ell & (-A^T(-\omega))^k \end{bmatrix}\begin{bmatrix} B \\ -Q(\omega) \end{bmatrix}$$

$$= Q^T(-\omega)A(\omega)^k B - B^T \sum_{\substack{i+\ell=k-1\\ i,\ell\geq o}} (-A^T(-\omega))^i P(\omega)/(j\omega)A(\omega)^\ell B \tag{4.6}$$

$$- B^T(-A^T(-\omega))^k Q(\omega).$$

Thus comparison yields the following expansion for $\Pi(\omega)$

$$\Pi(\omega) = \sum_{k=o}^{\infty} (j\omega)^{-k} R_k(\omega). \tag{4.7}$$

Lemma 4.2 For each $k \geq 0$ and each $\omega \in R_+$ one has

$$R_{2k}(\omega)^T = \overline{R_{2k}(\omega)} \quad \text{and} \quad R_{2k+1}(\omega)^T = -\overline{R_{2k+1}(\omega)},$$

i.e. R_{2k} is Hermitian and R_{2k+1} is skew-Hermitian.

Proof: Clearly the real matrix $R_o = R$ is symmetric. For $k \geq 1$, the proof follows by inspection of formula (4.6) and the properties of $A(\omega)$, $P(\omega)$, and $Q(\omega)$ mentioned above. □

Lemma 4.3 Suppose that for some $\ell \in \{0,1,2,...\}$ and all ω large enough, the following assumption holds:

$$R_k(\omega) = 0 \quad \text{for all} \quad 0 \leq k < \ell \quad \text{and} \quad R_\ell(\omega) \pm 0.$$

Then there exists $\omega_o > 0$ such that the following conditions are equivalent:

(i) There exists $\delta > 0$ for all $\omega \geq \omega_o$ there is $\eta \in R^m$ with $|\eta| = 1$ and $\eta^T\Pi(\omega)\eta < -\delta$;

(ii) There exists $\delta > 0$ such that for all $\omega \geq \omega_o$ there is $\eta \in R^m$ with $|\eta| = 1$ and $\eta^T j^{-\ell} R_\ell(\omega)\eta < -\delta/\omega^\ell$.

Proof: Suppose that (i) holds. Then for all $\omega \geq \omega_o$, there is $\eta \in R^m$ with

$$-\delta > \eta^T\Pi(\omega)\eta = \eta^T \sum_{k=\ell}^{\infty} (j\omega)^{-k} R_k(\omega)\eta$$

by (4.7) and assumption. But from (4.3), (4.6), and boundedness of

$|P(\omega)|$, $|Q(\omega)|$ it follows for all ω with $|\omega| > |L| + \varepsilon$ that

$$\sum_{k=\ell+1}^{\infty} |(j\omega)^{-k} R_k(\omega)|$$

$$\le |\omega|^{-1} \sum_{k=\ell+1}^{\infty} \{2|Q(\omega)|(|L|/|\omega|)^{k-1}|B| + |B|(|L|/|\omega|)^{k-1}|P(\omega)|\}$$

$$\le c|\omega|^{-\ell-1}(1-|L|/|\omega|)^{-1}$$

for some constant c which is independent of ω. Hence (ii) follows. The converse can be seen in the same way. □

Now one easily obtains the following *High-Frequency* Π-*Test*.

Theorem 4.4 Suppose that (x^o,u^o) is a local solution of Problem 1.2 (with $h = k \equiv 0$ and $\Omega = R^m$), and assume that conditions (3.5) and (4.1) are satisfied. Let $\ell \in \{0,1,2,\ldots\}$. Then either of the following conditions implies that (x^o,u^o) is locally proper:

(i) There exist $\omega_o > 0$ and $\delta > 0$ such that for all $\omega \ge \omega_o$ and all $k = 0,1,\ldots,2\ell-1$ one has $R_k(\omega) = 0$ and there exists $\eta \in R^m$ with $|\eta| = 1$ such that $(-1)^{\ell}\eta^T R_{2\ell}(\omega)\eta < -\delta/\eta^{2\ell}$.

(ii) There exist $\omega_o > 0$ and $\delta > 0$ such that for all $\omega \ge \omega_o$ and all $k = 0,1,\ldots,2\ell$ one has $R_k(\omega) = 0$ and there exists $\eta \in R^m$ with $|\eta| = 1$ such that $(-1)^{\ell+1} j\eta^T R_{2\ell+1}(\omega)\eta < \delta/\omega^{2\ell+1}$.

Proof: By Lemma 4.3, $\Pi(\omega)$ is partially negative iff $j^{-\ell}R_\ell(\omega)$ is. This together with Theorem 3.3 implies that (i) is sufficient for local properness.

The assertion for odd coefficients follows in the same way noting that

$$j^{-(2\ell+1)}R_{2\ell+1}(\omega) = (-1)^{\ell+1} jR_{2\ell+1}(\omega).$$

□

Consider again the special case (3.20) of a single constant delay in the system equation.

Here one obtains

$$R_o = R$$

$$R_k(\omega) = [Q(-\omega)^T \; B^T] \begin{bmatrix} A_0+A_1\exp(-j\omega r) & 0 \\ -P(\omega)/\omega & -A_0-A_1\exp(j\omega r) \end{bmatrix}^k \begin{bmatrix} B \\ -Q(\omega) \end{bmatrix}.$$

Comparing this with (3.24) one observes that now, for the computation of $R_k(\omega)$, it is not necessary to compute $T(\omega) = \Delta^{-1}(j\omega)B$, i.e. to invert $\{j\omega I-A_0-A_1\exp(-j\omega r)\}$.

Remark 4.5 A high-frequency π-Test for ordinary differential equations was first proposed by Watanabe/Nishimura/Matsubara [1976] (they called it "singular control test"). An application is given in Watanabe/Onogi/Matsubara [1981].

The following simple example illustrates usefulness of the High-Frequency π-Test.

Example 4.6 Minimize $1/\tau \int_0^\tau [x_1(t)^2-2x_2(t)^2+u(t)]dt$ s.t.

$$\dot{x}(t) = x_2(t)$$

$$\dot{x}_2(t) = -ax_2(t) - x_1(t-1) + u(t).$$

By a standard result, Hale [1977, Theorem A6], the stability condition (4.1) is satisfied provided that

$$a > \sin \xi/\xi,$$

where ξ is the unique root with $0 < \xi < \pi/2$ of the equation

$$\xi^2 = \cos \xi.$$

The corresponding steady steady state problem

$$\text{Minimize } x_1^2 - 2x_2^2 + u \text{ s.t.}$$

$$0 = x_2$$

$$0 = -x_1 + u$$

has $(x_1,x_2,u) = (-1/2,0,-1/2)$ as optimal solution. The High-Frequency π-Test, Theorem 4.4, applies, since

$$R_0 = H_{uu} = 0$$

and

$$jR_1(\omega) = -4/\omega.$$

Observe that for scalar controls the choice of η poses no problem; furthermore, the steady state Lagrange multipliers are not needed here, since the system equation is linear.

The problem is thus locally proper for all sufficiently high frequencies.

5. Strong Tests

The tests for local properness developed in Sections 3 and 4 above are based on (second order) *weak variations*. In this section we consider tests based on *strong variations*, i.e. global variations of the control. A "strong" test of this kind is furnished by the global maximum principle Theorem IV.2.1: In general, solutions of Problem 1.2 will not satisfy the global maximum principle for Problem 1.1. For ordinary differential equations this is well-known in the literature and has been treated in full detail with many examples and counterexamples in Gilbert [1977,1978]. We will not discuss this test, which, naturally, is ineffective, if f and g are affine-linear in u and Ω is convex.

Instead we will concentrate on the second order necessary optimality conditions derived in Section VI.4 for optimal relaxed solutions. This is justified in particular by Remark VI.2.6 which shows that often optimal ordinary solutions are also optimal among relaxed solutions. Recall $H(\varphi,\omega,y) = g(\varphi(0),\omega) - y^T f(\varphi,\omega)$.

For $(x^o,u^o) \in R^n \times rpm(\Omega)$ let

$$\Lambda := \{(1,y) \in R\times R^n : \mathcal{D}_1 H(\bar{x}^o,v^o,y) = 0 \qquad (5.1)$$
$$H(\bar{x}^o,v^o,y) = \min_{\omega\in\Omega} H(\bar{x}^o,\omega,y)\}.$$

Theorem 5.1 Suppose that $(x^o,v^o) \in R^n \times rpm(\Omega)$ is a local minimum of the Relaxed Problem 1.2 with $h = k \equiv 0$, and assume, in addition to (2.17) and (2.18).

The maps f and g are twice continuously Fréchet differentiable with respect ot the first argument and $|\mathcal{D}_1\mathcal{D}_1 f(\varphi,\omega)|$ and $|g_{xx}(x,\omega)|$ are bounded for bounded arguments. (5.2)

Then (x^o,v^o) is locally proper if there exist $(x,v) \in C(-r,\tau;R^n) \times S$ satisfying

$$\int_0^\tau [g_x(x^o,v^o)x(t) + g(x^o,v(t)-v^o)]dt \le 0 \qquad (5.3)$$

$$x_0 = x_\tau,\ \dot{x}(t) = \mathcal{D}_1 f(\bar{x}^o,v^o)x_t + f(\bar{x}^o,v(t)-v^o) \text{ a.a. } t \in [0,\tau] \qquad (5.4)$$

$$\max_{\lambda\in\Lambda} \int_0^{\tau} [\mathcal{D}_1\mathcal{D}_1 H(\bar{x}^0,v^0,\lambda)(x_t,x_t) + 2\mathcal{D}_1 H(\bar{x}^0,v(t)-v^0,\lambda)x_t]dt < 0.$$

Proof: This is a direct consequence of Theorem VI.4.5 and Proposition 2.8.

□

Remark 5.2 If f and g are affine linear in ω and Ω is convex, the assertion above reduces to Theorem 3.1.

CHAPTER VIII

A SCENARIO FOR LOCAL PROPERNESS

This chapter relates local properness to structural changes in the system equation. The analysis is motivated by the following consideration: Suppose a Hopf bifurcation occurs at $\alpha = \alpha_0$ in a system depending on a parameter $\alpha \in R$. If the generated periodic solution is "better" than the steady state solution, one will expect local properness at $\alpha = \alpha_0$. It turns out that under a controllability condition, this true *for all* α *close to* α_0.

The controllability condition guarantees that the free periodic motion can be approximated by forced periodic motions. In fact it is not necessary that actually a Hopf bifurcation occurs; instead the properties of what we call a *Controlled Hopf Bifurcation* are sufficient for local properness near α_0. Section 3 presents an example involving a retarded Liênard equation.

The main result of this chapter is Theorem 2.9.

1. Problem Formulation

In this chapter, we consider, together with its steady state version the following *parameter dependent* autonomous optimal periodic control problem.

Problem 1.1 Minimize $1/\tau \int_0^\tau g(x(s),u(s))ds$

$$\text{s.t.} \quad \dot{x}(t) = f(x_t,u(t),\alpha) \quad \text{a.a.} \quad t \in T := [0,\tau]$$

$$1/\tau \int_0^\tau k(x(t),u(t))dt = 0$$

$$x_0 = x_\tau$$

here $g: R^n \times R^m \to R$, $f = (f^i) : C(-r,0;R^n) \times R^m \times R \to R^n$

$k = (k^i) : R^n \times R^m \to R^{n_1}$; we admit control functions u in $L^\infty(0,\tau,R^m)$.

The corresponding steady state problem has the following form.

Problem 1.2$^\alpha$ Minimize $g(x,u)$

over $(x,u) \in R^n \times R^m$ s.t.

$$0 = f(\bar{x},u,\alpha)$$

$$0 = k(x,u)$$

where f, g, and k are as in Problem 1.1.

For fixed α, these are special cases of Problem VII.1.1 and Problem VII.1.2, respectively.

The following hypotheses are imposed throughout this chapter.

Hypothesis 1.3 The functions f, g, and k are twice continuously Fréchet differentiable and bounded together with their first and second Fréchet derivatives for bounded arguments.

Suppose that $(x^{\alpha_0}, u^{\alpha_0}) \in R^n \times R^m$ is a local minimum of Problem 1.2$^{\alpha_0}$. We will analyze if local minima (x^α, u^α) of Problem 1.2$^\alpha$ with α close to α_0 are locally proper. In view of Theorem II.3.3 we require a constraint qualification and a second order sufficient optimality condition in order to guarantee existence of (x^α, u^α). It is convenient, to write the latter condition using the Pontryagin function

$$H(\varphi,u,y,\alpha) := g(\varphi(0),u) + y^T \begin{pmatrix} f(\varphi,u,\alpha) \\ k(\varphi(0),u) \end{pmatrix} \tag{1.1}$$

where $(\varphi,u,y,\alpha) \in C(-r,0;R^n) \times R^m \times R^{n+n_1} \times R$.

Hypothesis 1.4 The gradients in $R^{n \times m}$

$$D_{1,2} f^j(\bar{x}^{\alpha_0}, u^{\alpha_0}), \quad j = 1,\dots,n,$$

$$k^j_{x,u}(x^{\alpha_0}, u^{\alpha_0}), \quad j = 1,\dots,n_1$$

are linearly independent.

Hypothesis 1.5 There exists $y^{\alpha_0} \in R^{n+n_1}$ such that

$$D_{1,2} H(\bar{x}^{\alpha_0}, u^{\alpha_0}, y^{\alpha_0}, \alpha_0) = 0 \tag{1.2}$$

and

$$D_{1,2} D_{1,2} H(\bar{x}^{\alpha_0}, u^{\alpha_0}, y^{\alpha_0}, \alpha_0)((\bar{x},u),(\bar{x},u)) > 0 \tag{1.3}$$

for all $(x,u) \in R^n \times R^m$ with

$$\mathcal{D}_{1,2}f(\bar{x}^{\alpha_0},u^{\alpha_0},\alpha_0)(x,u) = 0, \quad \dot{k}_{x,u}(x^{\alpha_0},u^{\alpha_0})(x,u) = 0. \tag{1.4}$$

2. Controlled Hopf Bifurcations

First we note the following immediate consequence of Theorem II.3.3.

Theorem 2.1 Suppose that Hypotheses 1.3 - 1.5 are satisfied for $(x^{\alpha_0},u^{\alpha_0}) \in R^n \times R^m$ satisfying the constraints of Problem 1.2^{α_0}.

Then

(i) $(x^{\alpha_0},u^{\alpha_0})$ is an isolated local minimum of Problem 1.2^{α_0} and the Lagrange multiplier y is determined uniquely by condition (1.2).

(ii) There exists a continuously differentiable function $\alpha \to (x(\alpha),u(\alpha),y(\alpha)) \in R^n \times R^m \times R^{n+n_1}$ defined on a neighborhood of α_0 such that

$$(x(\alpha_0),u(\alpha_0),y(\alpha_0)) = (x^{\alpha_0},u^{\alpha_0},y^{\alpha_0}) \tag{2.1}$$

and $(x^\alpha,u^\alpha,y^\alpha) := (x(\alpha),u(\alpha),y(\alpha))$ satisfy conditions (1.2) and (1.3) with α replaced for α^0; the points (x^α,u^α) are isolated local minima of Problem 1.2^α.

Remark 2.2 Colonius [1988] additionally allows control constraints of the form

$$q(u(t)) \in R^\ell_-, \quad q: R^m \to R^\ell \tag{2.2}$$

as well as state constraints in Problem 1.1 and still gets results analogous to those in the present section.

Define

$$L(\alpha) := \mathcal{D}_1 f(\bar{x}^\alpha,u^\alpha), \quad B(\alpha) = \mathcal{D}_2 f(\bar{x}^\alpha,u^\alpha). \tag{2.3}$$

The characteristic function of the linearized equation

$$\dot{x}(t) = L(\alpha)x_t, \quad t \in R \tag{2.4}$$

is given by

$$\Delta(z,\alpha) = zI - L(\alpha)(e^{z\cdot}I), \quad z \in C.$$

Lemma 2.3 Suppose that

$$\operatorname{rank} \Delta(j\omega_0,\alpha_0) = n-1 \quad \text{and} \tag{2.5}$$

$$\operatorname{rank} \Delta(j\omega,\alpha) = n \quad \text{for} \quad (\omega,\alpha) \neq (\omega_0,\alpha) \quad \text{close to} \quad (\omega_0,\alpha_0).$$

Then for α in a neighborhood of α_0, equation (2.3) has a simple eigenvalue $z(\alpha)$ with $z(\alpha_0) = j\omega_0$ and $z(\alpha)$ has a continuous derivative $z'(\alpha_0)$.

Proof: By Theorem 2.1, the map $\alpha \to L(\alpha)$ is continuously Fréchet differentiable and Hale [1977, Lemma 2.2, p.171] implies the assertion. □

Remark 2.4 Condition (2.5) does not require that an eigenvalue actually crosses the imaginary axis. However, (2.5) is valid, if a Hopf bifurcation occurs at $\alpha = \alpha_0$ (cp. Theorem III.3.3).

Lemma 2.5 Condition (2.5) implies that there exist a non-trivial τ-periodic solution p of equation (2.4) with $\alpha = \alpha_0$, $\tau := 2\pi/\omega_0$ and $p_1 \in C^n$ such that for every such p

$$p(t) = 2\gamma \operatorname{Re}(e^{j\omega_0 t} p_1), \quad t \geq 0 \tag{2.6}$$

for some $\gamma \in R$.

Proof: By assumption, the eigenspace corresponding to $z = j\omega_0$ is one dimensional and the assertion follows (cp. Hale [1977]).

Lemma 2.6 Suppose that condition (2.5) is satisfied. Then the following two conditions are equivalent:

$$\text{There exists } \nu \in C^m \text{ with } p_1 = [\operatorname{Adj} \Delta(j\omega_0,\alpha_0)]B(\alpha_0)\nu, \tag{2.7}$$

where p_1 is given by Lemma 2.5;
the adjoint of $\Delta(j\omega_0,\alpha_0)$ satisfies

$$[\operatorname{Adj} \Delta(j\omega_0,\alpha_0)]B(\alpha_0) \neq 0. \tag{2.8}$$

Proof: Observe that

$$\Delta(j\omega_0,\alpha_0) \operatorname{Adj} \Delta(j\omega_0,\alpha_0) = \det \Delta(j\omega_0,\alpha_0)\cdot I$$

(see e.g. Kowalsky [1963, Kapitel 4]). Thus the range of

$$[\operatorname{Adj} \Delta(j\omega_0,\alpha_0)]B(\alpha_0)$$

is always contained in the kernel of $\Delta(j\omega_0,\alpha_0)$, which is spanned by p_1. □

Condition (2.7) may be viewed as a "controllability condition" for the periodic solution (2.6) corresponding to p_1.

The Pontryagin function H for Problem 1.1 has been defined in (1.1). We abbreviate

$$P(\omega,\alpha) := \mathcal{D}_1\mathcal{D}_1 H(\bar{x}^{\alpha},u^{\alpha},y^{\alpha},\alpha)(e^{-j\omega\cdot}I,e^{j\omega\cdot}I) \in C^{n\times n}; \tag{2.9}$$

similarly for $Q(\omega,\alpha)$ and $R(\alpha)$ (comparing this with (VII.3.15) we have supressed dependence on λ, since λ is unique here by Proposition VII.2.6 and Theorem 2.1).

Lemma 2.7 Suppose that condition (2.5) is satisfied. Then

$$\bar{p}_1^T\ P(\omega_0,\alpha_0)p_1 = \int_0^{\tau} \mathcal{D}_{1,2}\mathcal{D}_{1,2}H(\bar{x}^{\alpha_0},u^{\alpha_0},y^{\alpha_0},\alpha_0)((p_t,0),(p_t,0))dt$$

for p_1,p as in Lemma 2.5.

Proof: Obvious from the definitions and Lemma 2.5. □

This lemma shows that the condition

$$\bar{p}_1^T\ P(\omega_0,\alpha_0)p_1 < 0 \tag{2.10}$$

may be viewed as a second order "properness condition" for the periodic solution (2.6) corresponding to p_1.

Next we introduce the central notion of this chapter.

Definition 2.8 Let $(x^{\alpha_0},u^{\alpha_0}) \in R^n\times R^m$ satisfy the constraints of Problem 1.2^{α_0}. A *Controlled Hopf Bifurcation* with frequency ω_0 occurs at $\alpha = \alpha_0$ if Hypotheses 1.3 - 1.5 hold and conditions (2.5) and (2.7) are satisfied. Define

$$\begin{aligned} C^{m\times m} \ni \Pi(\omega,\alpha) := {} & B(\alpha)^T\Delta^{-1}(-j\omega,\alpha)^T P(\omega,\alpha)\Delta^{-1}(j\omega,\alpha)B(\alpha) \\ & + B(\alpha)^T\Delta^{-1}(j\omega,\alpha)^T Q(\omega,\alpha) \\ & + Q(-\omega,\alpha)^T\Delta^{-1}(j\omega,\alpha)B(\alpha) + R(\alpha). \end{aligned} \tag{2.11}$$

Condition (2.5) implies that $\Pi(\omega,\alpha)$ exists for all $(\omega,\alpha) \neq (\omega_0,\alpha_0)$ in a neighborhood of (ω_0,α_0).

Comparing (2.12) with (VII.3.16), we have supressed dependence of Π on the Lagrange multiplier $\lambda = (\lambda_0,y)$, since the Lagrange multipliers for Problem 1.1^{α}, $\alpha \neq \alpha_0$, are unique.

Now we can state the main result of this chapter.

Theorem 2.9 Assume that in Problem 1.1 a controlled Hopf bifurcation with frequency ω_0 occurs at $\alpha = \alpha_0$.

If the properness condition (2.10) is satisfied then there exists a neighborhood N of (ω_0,α_0) such that the steady states (x^α,u^α) which are isolated local minima of Problem 1.2^α are locally proper and

$$\bar{\nu}^T \Pi(\omega,\alpha)\nu < 0 \quad \text{for all} \quad (\omega,\alpha) \in N, \quad (\omega,\alpha) \neq (\omega_0,\alpha_0). \tag{2.12}$$

where ν is given by (2.7).

Proof: Existence of isolated local minima (x^α,u^α) follows from Theorem 2.1. Furthermore, inequality (2.12) implies by Theorem VII.3.3 the asserted local properness. Hence it remains to establish (2.12).
By (2.10) and (2.7)

$$\begin{aligned} 0 &> \bar{p}_1^T \, P(\omega_0,\alpha_0)p_1 \\ &= \bar{\nu}^T \, B(\alpha_0)^T[\mathrm{Adj}\ \Delta(-j\omega_0,\alpha_0)]^T \, P(\omega_0,\alpha_0)[\mathrm{Adj}\ \Delta(j\omega_0,\alpha_0)]B(\alpha_0)\nu. \end{aligned} \tag{2.13}$$

But $\mathrm{Adj}\ \Delta(j\omega,\alpha)$ and $P(\omega,\alpha)$ are continuous with respect to (ω,α). Furthermore

$$[\det \Delta(j\omega,\alpha)]^2 > 0$$

for $(\omega,\alpha) \neq (\omega_0,\alpha_0)$ in a neighborhood of (ω_0,α_0).
Hence in a neighborhood of (ω_0,α_0)

$$\begin{aligned} &[\det \Delta(j\omega,\alpha)]^{-2}\{\bar{\nu}^T B(\alpha)^T[\mathrm{Adj}\ \Delta(-j\omega,\alpha)]^T P(\omega,\alpha)[\mathrm{Adj}\ \Delta(j\omega,\alpha)]B(\alpha)\nu\} \\ &= \nu^T B(\alpha)^T \Delta^{-1}(-j\omega,\alpha)^T P(\omega,\alpha)\Delta^{-1}(j\omega,\alpha)B(\alpha)\nu < 0 \end{aligned} \tag{2.14}$$

For $(\omega,\alpha) \to (\omega_0,\alpha_0)$, $\det \Delta(j\omega,\alpha)$ converges to zero, while the second factor $\{\ldots\}$ converges to $p_1^T H_{xx}(\omega_0,\alpha_0)p_1 \neq 0$.

Now consider the definition (2.11) of $\Pi(\omega,\alpha)$:

For $(\omega,\alpha) \to (\omega_0,\alpha_0)$ the first summand tends to infinity with the square of $\det [\Delta(j\omega,\alpha)]^{-1}$, the other tend to infinity at most with $\det [\Delta(j\omega,\alpha)]^{-1}$. Thus the first summand becomes dominant and by (2.14)

$$\bar{\nu}^T \, \Pi(\omega,\alpha)\nu < 0$$

for all (ω,α) in a neighborhood of (ω_0,α_0), $(\omega,\alpha) \neq (\omega_0,\alpha_0)$.
□

Remark 2.10 The second order sufficiency condition (Hypothesis 1.5) for the steady state Problem 1.2 is needed in order to guarantee smooth dependence of $(x^\alpha,u^\alpha,y^\alpha)$ on α. If this can be guaranteed by other arguments (e.g. if the steady state problem is independent of α as in the example in Section 3, below) we can replace Hypothesis 1.5 by the assumption that (x^α,u^α) are a local minimum of Problem 1.2^α.

Remark 2.11 Obviously, Theorem 2.9 remains valid if the parameter α varies only in an open interval. However, it also remains valid, if α varies in a *closed* interval, provided that existence and smooth dependence of $(x^\alpha,u^\alpha,y^\alpha)$ is guaranteed.

We have the following partial converse of Theorem 2.9.

Theorem 2.12 Assume that in Problem 1.1 a controlled Hopf bifurcation with frequency ω_0 occurs at $\alpha = \alpha_0$.
If there exists a sequence $(\omega_n,\alpha_n) \to (\omega_0,\alpha_0)$, $(\omega_n,\alpha_n) \neq (\omega_0,\alpha_0)$, with

$$\bar{\nu}^T \, \Pi(\omega_n,\alpha_n)\nu > 0 \tag{2.15}$$

then the properness condition (2.10) is violated.

Proof: Conditions (2.15) and (2.7) imply

$$0 < \bar{\nu}^T \, \Pi(\omega_n,\alpha_n)\nu = \bar{\nu}^T \, B(\alpha_n)^T \Delta^{-1}(-j\omega_n,\alpha_n)^T P(\omega_n,\alpha_n)\Delta^{-1}(j\omega_n,\alpha_n)B(\alpha_n)\nu$$
$$+ \bar{\nu}^T \{B(\alpha_n)^T \Delta^{-1}(-j\omega_n,\alpha_n)^T Q(\omega_n,\alpha_n)$$
$$+ Q(-\omega_n,\alpha_n)^T \Delta^{-1}(j\omega_n,\alpha_n)B(\alpha_n) + H_{uu}(\alpha_n)\}\nu.$$

The first summand equals

$$[\det \Delta^{-1}(-j\omega_n,\alpha_n)]^{-2}[\bar{\nu}^T B(\alpha_n)^T [\operatorname{Adj} \Delta(-j\omega_n,\alpha_n)]^T$$
$$P(\omega_n,\alpha_n) \operatorname{Adj} \Delta(j\omega_n,\alpha_n)B(\alpha_n)\nu].$$

Again $[\det \Delta(-j\omega_n,\alpha_n)]^2 > 0$, and the second factor converges to

$$\nu^T \, B(\alpha_0)[\operatorname{Adj} \Delta(-j\omega_0,\alpha_0)]^T P(\omega_0,\alpha_0) \operatorname{Adj} \Delta(j\omega_0,\alpha_0)B(\alpha_0)\nu = \bar{p}_1^T \, P(\omega_0,\alpha_0)p_1.$$

Arguing as in the proof of Theorem 2.9, we obtain

$$\bar{p}_1^T \, P(\omega_0,\alpha_0)p_1 > 0. \tag{2.17}$$

□

Remark 2.13 Suppose that a Hopf bifurcation occurs at $\alpha = \alpha_0$. Then Theorem 2.9 may be interpreted as follows: At $\alpha = \alpha_0$, a "natural" periodic solution of $\dot{x}(t) = f(x_t,u^\alpha,\alpha)$ bifurcates from the steady state x^{α}, $\alpha = \alpha_0$. By (2.10), this periodic motion shows better average performance than the steady state. Condition (2.5) is satisfied and the controllability condition (2.7) guarantees that (by continuity) for all α near α_0 the periodic trajectory can be approximated by trajectories corresponding to a periodic control. Hence, for α near α_0, the points x^{α} are locally proper. Suppose e.g. non trivial periodic trajectories exist for $\alpha > \alpha_0$. Then, also for $\alpha < \alpha_0$, where no free periodic trajectory exists, one can generate periodic trajectories by appropriate periodic controls. Thus it is not surprising that the assumption can be weakened by requiring only the existence of a controlled Hopf bifurcation as defined in Definition 2.8: It is not necessary that the nonlinear equation actually has a free periodic trajectory.

Remark 2.14 Russell [1982] observed another connection between Hopf bifurcation and optimal periodic control. He was interested in coupled nonlinear regulators where a Hopf bifurcation causes periodic motions which he wanted to dampen. He considered this as an optimal periodic control problem where the performance criterion is constructed in such a way as to minimize the amplitude of the oscillators.

Remark 2.15 The results in this section have been developed under the assumption that $j\omega_0$ is a simple eigenvalue of (2.4). This, though facilitating the arguments, does not appear crucial for the arguments in the proof of Theorem 2.7.

Remark 2.16 The stability properties of the forced equations near $\alpha = \alpha_0$ may be very complicated; cp. Gambaudo [1985] for a classification in the case of two dimensional ordinary differential equations.

Remark 2.17 Lorentz [1978], Vogel [1979] establish a connection, different from that in the present chapter, between normality of Lagrange multipliers and bifurcation. Another direction of research connects constraint qualifications for finite dimensional optimization problems with bifurcations in the solution sets (Kojima [1980], Jongen/Jonker/Twilt [1983]).

3. Example: Periodic Control of Retarded Liénard Equations

We consider the following optimal periodic control problem and its steady state version.

<u>Problem 3.1</u> Minimize $1/\tau \int_0^\tau [1/2\, u(s)^2 - x(s)]ds$ s.t.

$$\ddot{x}(t) + f(x(t))\dot{x}(t) + g(x(t-r)) = u(t), \quad \text{a.a. } t \in [0,\tau] \tag{3.1}$$

$$x_0 = x_\tau, \quad \dot{x}_0 = \dot{x}_\tau \tag{3.2}$$

$$\int_0^\tau u(t)dt = 0 \tag{3.3}$$

where $x(t), u(t) \in R$, and we assume

$f,g: R \rightarrow R$ are twice continuously differentiable in a neighborhood of zero with (3.4)

$f(0) = g(0) = 0, \quad g'(0) = 1, g''(0) = -1$

$g(x) \neq 0$ for $x > 0$.

A typical example of g is sketched in Fig. 1.

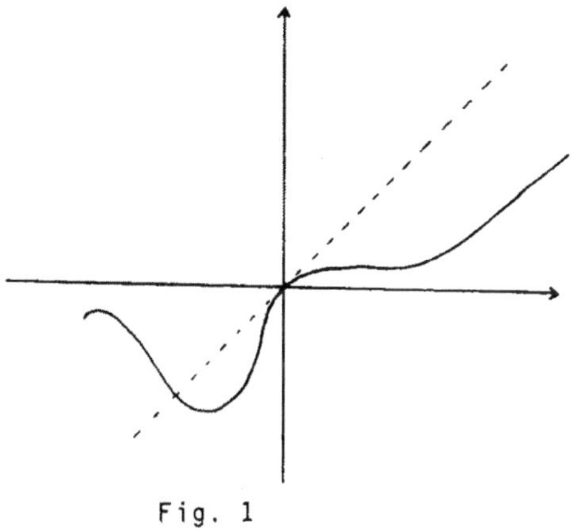

Fig. 1

Naturally, the system equation (3.1) is equivalent to a two-dimensional system of first order equations. Note that the time transformation $t := tr$, for $r > 0$, yields

$$\dot{x}_1(t) = 1/r\, x_2(t) \tag{3.5}$$
$$\dot{x}_2(t) = 1/r\, [-f(x_1(t))x_2(t) - g(x(t-1)) + u(t)]$$

and we may consider $\alpha = r$ as a bifurcation parameter.

The steady state problem corresponding to Problem 3.1 is

Problem 3.2 Minimize $1/2\ u^2 - x_1$

$$\text{s.t.}\quad 0 = x_2$$

$$0 = -f(x_1)x_2 - g(x_1) + u$$

$$0 = u\ .$$

Assumption (3.4) guarantees that $(x^0,u^o) = (0,0) \in R^2 \times R$ is the unique minimum of Problem 3.2.

Note that Problem 3.2 is independent of α, hence Remark 2.10 applies and we do not need Hypothesis 1.5. Concerning Hypothesis 1.4, linearization of the constraints in Problem 3.2 yields the matrix

$$\begin{bmatrix} 0 & 1 & 0 \\ -f'(x_1)x_2 - g'(x_1) & -f(x_1) & 1 \\ 0 & 0 & 1 \end{bmatrix}$$

and, evaluated at zero

$$\begin{bmatrix} 0 & 1 & 0 \\ -1 & 0 & 1 \\ 0 & 0 & 1 \end{bmatrix} \tag{3.6}$$

The matrix (3.6) has full rank. Thus Hypothesis 1.4 is valid here.

The Pontryagin function $H: R^2 \times R^2 \times R \times R^3 \times (0,\infty) \to R$ is given by (cp. (VII.3.22))

$$H(x,y,u,\lambda,\tau) = 1/2\ u^2 - x_1 + \lambda^T \begin{pmatrix} x_2 \\ f(x_1)x_2 - g(y_1) + u \\ u \end{pmatrix}$$

$$= 1/2\ u^2 - x_1 + \lambda_1 x_2 - \lambda_2 f(x_1)x_2 - \lambda_2 g(y_1) + \lambda_2 u + \lambda_3 u$$

(here the delayed variables are denoted by $y = (y_1,y_2)$ and $\lambda = (\lambda_1,\lambda_2,\lambda_3)^T$).

There exists a Lagrange multiplier $\lambda \in R^3$ for the steady state Problem 3.2 satisfying

$$(-1\ \ 0\ \ 0) + \lambda^T \begin{pmatrix} 0 & 1 & 0 \\ -1 & 0 & 1 \\ 0 & 0 & 1 \end{pmatrix} = 0$$

i.e.

$$\lambda_1 = 0, \quad \lambda_2 = -1, \quad \lambda_3 = 1. \tag{3.7}$$

Linearizing the system equation (3.1) around $(x_1^0, x_2^0, u^0) = (0,0,0)$, we find

$$\dot{x}(t) = A_0 x(t) + A_1 x(t-r) + Bu(t) \tag{3.8}$$

with

$$A_0 := \begin{pmatrix} 0 & 1 \\ 0 & 0 \end{pmatrix}, \quad A_1 := \begin{pmatrix} 0 & 0 \\ -1 & 0 \end{pmatrix}, \quad B := \begin{pmatrix} 0 \\ 1 \end{pmatrix};$$

equivalently

$$\ddot{x}(t) + x(t-r) = u(t). \tag{3.9}$$

Thus

$$\Delta(z,r) = zI - A_0 - A_1 \exp(-zr) = \begin{pmatrix} z & -1 \\ \exp(-zr) & z \end{pmatrix} \tag{3.10}$$

and the characteristic equation is

$$\det \Delta(z,r) = z^2 + \exp(-zr) = 0. \tag{3.11}$$

Lemma 3.3 (i) There exists an eigenvalue z on the imaginary axis iff $r = r_n = 2n\pi$, $n = 0,1,2,\ldots$

(ii) For $r = r_n$ the eigenvalues on the imaginary axis are $z = \pm j$

(iii) For r in a neighborhood of r_n, $r = 1,2,\ldots$, there exists a continuously differentiable function $r \to z(r)$ such that $z(r)$ is a simple eigenvalue, $z(r_n) = jr_n$ and $\operatorname{Re} z'(r_n) > 0$.

Proof: Follows by an elementary analysis of (3.11). □

Remark 3.4 For $n = 0$ a function $z(r)$ with the properties above exists on the intersection of $[0,\infty)$ with a neighborhood of $r_0 = 0$.

Remark 3.5 The lemma implies that in equation (3.5) Hopf bifurcations occur at $\tilde{r}_n := 1/2n\pi$, $n = 1,2,\ldots$ Here the eigenvalues at the imaginary axis are $z = \pm j/\tilde{r}_n = \pm 2n\pi j$. This follows, since the characteristic equation of the corresponding linearized equation is

$$\det \tilde{\Delta}(z,r) = 1/r[r^2 z^2 + \exp(-z)] = 0.$$

It is not clear, if also at $n = 0$ a Hopf bifurcation occurs, since this case is not covered by the standard theorem on Hopf bifurcation for delay equations (cp. Hale [1977]).

A nontrivial periodic solution of (3.8) with $r = r_n$, $n = 0,1,2,\dots$, having period $\tau = 2\pi$ is given by

$$p(t) = 2\begin{pmatrix} \cos t \\ -\sin t \end{pmatrix} \tag{3.12}$$

which has Fourier coefficients

$$p_1 = \hat{p}(1) = \begin{pmatrix} 1 \\ j \end{pmatrix}, \quad \bar{p}_1 = \hat{p}(-1) = \begin{pmatrix} 1 \\ -j \end{pmatrix}, \quad \hat{p}(k) = 0 \text{ for } k \neq \pm 1.$$

In fact, controlled Hopf bifurcations occur at $r = r_n$, $n = 0,1,2,\dots$, since condition (2.8) is satisfied:

$$\mathrm{Adj}\ \Delta(j\omega_n, r_n)B = \begin{pmatrix} j\omega_n & 1 \\ -\exp(-j\omega_n r) & j\omega_n \end{pmatrix}\begin{pmatrix} 0 \\ 1 \end{pmatrix} = \begin{pmatrix} 1 \\ j\omega_n \end{pmatrix} \neq 0 .$$

Finally, we verify the properness condition (2.10).
Here (cp. (VII.3.23)) one has

$$P(\omega) = H_{xx} + 2H_{xy}\exp(-j\omega r) + H_{yy}$$

and

$$H_{xy} = 0$$

$$H_{xx} = \begin{pmatrix} 0 & f'(0) \\ f'(0) & 0 \end{pmatrix}$$

$$H_{yy} = \begin{pmatrix} g''(0) & 0 \\ 0 & 0 \end{pmatrix} = \begin{pmatrix} -1 & 0 \\ 0 & 0 \end{pmatrix}$$

Thus

$$P(\omega) = \begin{pmatrix} -1 & f'(0) \\ f'(0) & 0 \end{pmatrix}, \tag{3.13}$$

and

$$\bar{p}^T P(\omega)p_1 = (1 \quad -j)\begin{pmatrix} -1 & f'(0) \\ f'(0) & 0 \end{pmatrix}\begin{pmatrix} 1 \\ j \end{pmatrix} = -1 < 0. \tag{3.14}$$

Thus the properness condition (2.10) is verified.

We have shown that in equation (3.1) for $n = 0,1,2,\ldots$ controlled Hopf bifurcations occur satisfying the properness condition (2.10).

Hence Theorem 2.9 and Remark 2.11 imply that $(x^0,u^0) = (0,0) \in R^2 \times R$ is locally proper for Problem 3.1 with (ω,r) in a neighborhood of $(\omega_n,r_n) = (1,2n\pi)$, $(\omega,r) \neq (\omega_n,r_n)$ for all $n = 0,1,2,\ldots$

Now we compute $\Pi(\omega,r)$ in order to see how large these neighborhood are. One has

$$Q(\omega) = H_{ux} + H_{uy}\exp(-j\omega r), \quad R = H_{uu}$$

and

$$H_{ux} = 0, \quad H_{uy} = 0, \quad H_{uu} = 1.$$

Thus by (VII.3.24)

$$\Pi(\omega,r) = B^T\Delta^{-1}(-j\omega,r)^T P(\omega,r)\Delta^{-1}(j\omega,r)B + 1. \tag{3.15}$$

by (3.10)

$$\Delta^{-1}(z,r)B = [\det \Delta(z,r)]^{-1} \begin{pmatrix} z & 1 \\ -\exp(zr) & z \end{pmatrix}\begin{pmatrix} 0 \\ 1 \end{pmatrix}$$

$$= [\det \Delta(z,r)]^{-1} \begin{pmatrix} 1 \\ z \end{pmatrix}$$

and by (3.11)

$$\det \Delta(-j\omega,r)\cdot\det \Delta(j\omega,r) = \omega^4 - 2\omega^2 \cos^2(\omega r) + 1.$$

Thus (3.15) and (3.13) yield

$$\Pi(\omega,r) = [1 \quad -j\omega]\begin{pmatrix} -1 & f'(0) \\ f'(0) & 0 \end{pmatrix}\begin{pmatrix} 1 \\ j\omega \end{pmatrix}(\omega^4 - 2\omega^2 \cos^2(\omega r) + 1)^{-1} + 1 \tag{3.16}$$

$$= 1 - 1/[\omega^4 - 2\omega^2 \cos^2(\omega r) + 1].$$

A simple analysis of the function Π yields that for all $r,\omega \in R_+$, $\omega \neq 1$

$$1 - 1/(\omega^2-1)^2 \le \Pi(\omega,r) \le 1 - 1/(\omega^4+1)$$

$$\Pi(0,r) = 0; \quad \lim_{\omega\to\infty} \Pi(\omega,r) = 1.$$

Figures 2 - 5 show $\Pi(\omega,r)$ for different values of r (here $X = \omega$, $Y = r$, $Z = \Pi(\omega,r)$). (I acknowledge use of programs by Dr. M. Pratt for the plotting of the diagrams.)

A significant feature of this example is that the zones of local properness (including the ω-intervals where $\Pi(\omega,r) < 0$) which occur near a Hopf bifurcation at $r = r_n$ (indicated by a negative pole of $\Pi(\omega,r_n)$ at $\omega = 1$) do not vanish for increasing r. Thus for large delays r, $\Pi(\omega,r)$ becomes very oscillatory (see Figure 5).

Remark 3.6 It is easy to check that the function $\Pi: R_+ \times R_+ \to R \cup \{-\infty\}$ has no local minima besides $(\omega,r) = (1,r_n)$. This may be interpreted in the following way: Local properness in this problem occurs *only* via the mechanism described by Corollary 2.9. Naturally, this may not be true for other problems (e.g. local properness may be due to non-linearities in the performance criterion).

Remark 3.7 Already the unretarded equation (3.1) with $r = 0$ shows a highly complex behaviour under periodic excitations $u(\cdot)$. For the retarded case we cite the following result from DePascale/Iannacci [1983,Theorem 4]:

Let $f,g: R \to R$ be continuous and assume that one of the following conditions is satisfied:

(i) there exists $m > 0$ such that for $|y| \geq m$ one has $yg(y) \leq 0$ and $\liminf_{|y|\to\infty} g(y)/y > -1$

(ii) there exists $m > 0$ such that for $|y| \geq m$ one has $yg(y) \geq 0$ and $\liminf_{|y|\to\infty} g(y)/y < 1$.

Then for every $r \in [0,2\pi]$, equation (3.1) has at least one 2π-periodic solution for every $u \in L^1_{2\pi}$ with

$$\int_0^{2\pi} u(t)dt = 0.$$

Remark 3.8 Optimal control problems for unretarded Liênard equations (time optimal control to the origin) are surveyed in Barbanti [1980].

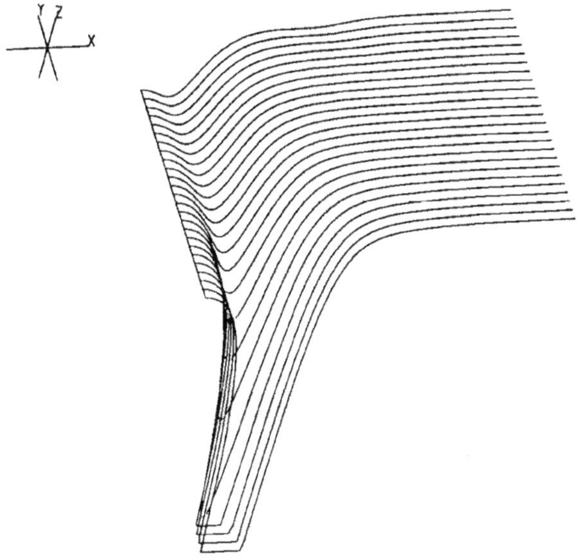

Fig. 2 shows $\Pi(\omega,r)$, $0 \le \omega \le 4$, for different values of r between $r = 0$ and $r = 3$ ($X = \omega$, $Y = r$, $Z = \Pi(r,\omega)$). The function values are cut off for $z < -3$.

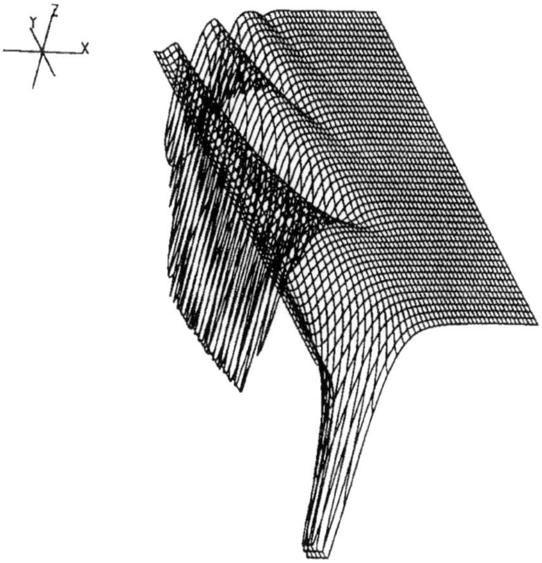

Fig. 3 shows $\Pi(\omega,r)$, $0 \le \omega \le 4$, for different values of r between $r = 0$ and $r = 10$.

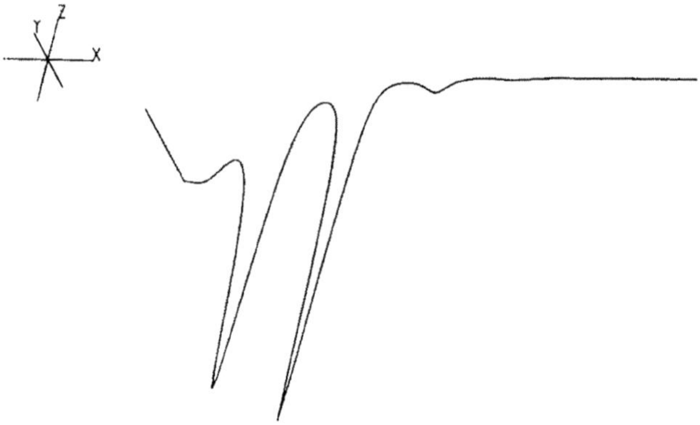

Fig. 4 shows $\Pi(\omega,r)$, $0 \le \omega \le 4$, for $r = 10$.

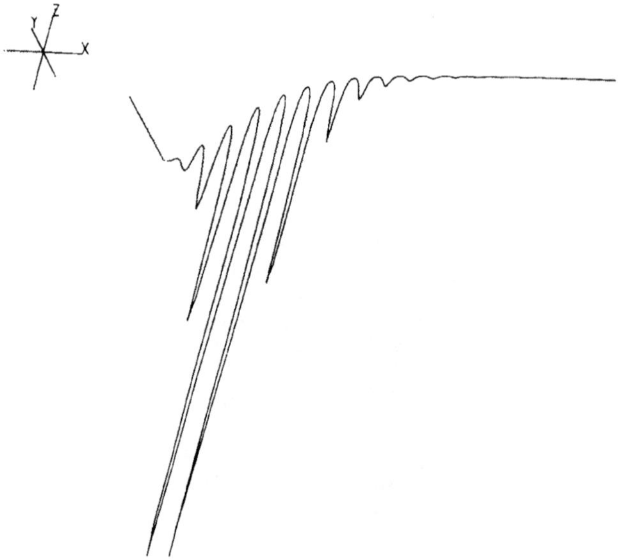

Fig. 5 shows $\Pi(\omega,r)$, $0 \le \omega \le 4$, for $r = 30$.

CHAPTER IX

OPTIMAL PERIODIC CONTROL OF ORDINARY DIFFERENTIAL EQUATIONS

In this chapter the results of the preceding chapters are specialized to the case of ordinary differential equations. We restrict our attention to second order necessary optimality conditions for weak local minima, Section 2, and discuss in Section 3 a Π-Test based on these conditions; the novel feature here is the inclusion of state constraints. The final Section 4 analyses local properness near a controlled Hopf bifurcation in a simple model of a continuous flow stirred tank reactor (CSTR).

1. Problem Formulation

We consider the following autonomous optimal periodic control problem for ordinary differential equations under state and control constraints and an isoperimetric constraint.

Problem 1.1 Minimize $1/\tau \int_0^\tau g(x(t),u(t))dt$

$$\text{s.t.} \quad \dot{x}(t) = f(x(t),u(t)) \quad \text{a.a.} \quad t \in T:= [0,\tau]$$

$$x(0) = x(\tau)$$

$$h(x(t)) \in R^\ell_-$$

$$u(t) \in \Omega$$

$$\int_0^\tau k(x(t),u(t))dt = 0;$$

here $g: R^n \times R^m \to R$, $f: R^n \times R^m \to R^n$, $h: R^n \to R^\ell$, $k: R^n \times R^m \to R^n$ and $\Omega \subset R^m$.

We allow control functions u in

$$\mathcal{U}_{ad} := \{u \in L^\infty(T;R^m) : u(t) \in \Omega \text{ a.e.}\}.$$

This is a special case of Problem VII.1.1. The corresponding steady state problem is

Problem 1.2 Minimize $g(x,u)$ over $(x,u) \in R^n \times R^m$ s.t.

$$f(x,u) = 0$$
$$h(x) \in R^{\ell}_{-}$$
$$u \in \Omega$$
$$k(x,u) = 0,$$

where f,g,h,k, and Ω are as Problem 1.1.

Notions for optimality and local properness in these problems are as those in Chapters IV - VII. We impose throughout the following hypothesis.

Hypothesis 1.3 The functions f,g,h and k are twice continuously differentiable; the set Ω is closed and convex.

Let $(x^0,u^0) \in C(T,R^n) \times L^\infty(T;R^m)$ be a weak local minimum of Problem 1.1.

2. Necessary Optimality Conditions

We omit explicit statement of optimality conditions for strong local minima (i.e. a global maximum principle) and of first order conditions for weak local and local relaxed minima. Instead we concentrate on second order conditions for weak local minima.

Define the attainability cone A for Problem 1.1 as

$$\begin{aligned} A := \{(x(\tau)-x(0),z) \in R^n \times R^{n_1}: &\text{ there exist } \alpha > 0 \text{ and} \\ u \in L^\infty(T;R^m) &\text{ with } u^0(t) + \alpha u(t) \in \Omega \text{ a.e. s.t.} \\ \dot{x}(t) &= f_x(x^0(t),u^0(t))x(t) + f_u(x^0(t),u^0(t))u(t) \text{ a.a. } t \in T, \\ z &= \int_0^\tau [k_x(x^0(t),u^0(t))x(t) + k_u(x^0(t),u^0(t))u(t)]dt\}. \end{aligned} \tag{2.1}$$

We formulate the following constraint qualification.

$$\begin{aligned} &\text{There exist } \tilde{u} \in L^\infty(T;R^m) \text{ and } \tilde{x} \in C(T;R^m) \text{ s.t.} \\ &\text{for some } \alpha > 0 \text{ one has } u^0(t) + \alpha u(t) \in \Omega \text{ a.e.,} \\ &x(0) = x(\tau), \\ &\dot{x}(t) = f_x(x^0(t)),u^0(t))x(t) + f_u(x^0(t),u^0(t))u(t) \text{ a.a. } t \in T, \\ &h_x(x^0(t))x(t) \in \text{int } R^{\ell}_{-} \text{ a. } t \in T, \\ &\int_0^\tau [k_x(x^0(t),u^0(t))x(t) + k_u(x^0(t),u^0(t))u(t)]dt = 0. \end{aligned} \tag{2.2}$$

For $x \in R^n$, $u \in R^m$, $\lambda = (\lambda_0, y, y_1, \mu) \in R \times R^n \times R^{n_1} \times C(T;R^\ell)^*$ let, formally, the Pontryagin "function" H for Problem 1.1 be given by

$$H(x,u,\lambda) := \lambda_0 g(x,u) + y^T f(x,u) + y_1^T k(x,u) + d\mu^T h(x). \tag{2.3}$$

Then we can formulate the following second order necessary optimality conditions.

<u>Theorem 2.1</u> Suppose that $(x^0,u^0) \in C(T;R^n) \times L^\infty(T;R^m)$ is a weak local minimum of Problem 1.1 and assume that Hypothesis 1.3 holds.

Let $(x,u) \in C(T;R^n) \times L^\infty(T;R^m)$ satisfy

$$\int_0^\tau [g_x(x^0(t),u^0(t))x(t) + g_u(x^0(t),u^0(t))u(t)]dt \le 0 \tag{2.4}$$

$$x(0) = x(\tau), \; \dot{x}(t) = f_x(x^0(t),u^0(t))x(t) + f_u(x^0(t),u^0(t))u(t) \quad \text{a.a. } t \in T \tag{2.5}$$

$$h(x^0(t)) + h_x(x^0(t))x(t) \in \text{int } R^\ell_- \tag{2.6}$$

$$u^0(t) + \alpha u(t) \in \Omega \quad \text{for a.a. } t \in T \text{ and some } \alpha > 0 \tag{2.7}$$

$$\int_0^\tau [k_x(x^0(t),u^0(t))x(t) + k_u(x^0(t),u^0(t))u(t)]dt = 0 \tag{2.8}$$

Then there exist $\lambda_0 \ge 0$, $y_1 \in R^{n_1}$, non negative regular Borel measures μ_i on T supported on the sets $\{t: h^i(x^0(t)) = 0\} \cap \{t: h^i_x(x^0(t))x(t) = 0\}$, $i = 1,\dots,\ell$ and a τ-periodic solution y of the adjoint equation

$$y(s) = y(\tau) + \int_s^\tau H_x(x^0(t),u^0(t),\lambda(t))dt, \quad t \in T, \tag{2.9}$$

where $\lambda(t) = (\lambda_0, y(t), z, \mu) \in R \times R^n \times R^{n_1} \times C(T;R^\ell)^*$ with $\lambda(\tau) \neq 0$ and

$$H_u(x^0(t),u^0(t),\lambda(t))[\omega - u^0(t)] \ge 0 \quad \text{for all } \omega \in \Omega \text{ and a.a. } t \in T \tag{2.10}$$

$$\begin{aligned}\int_0^\tau \{ & x^T(t)H_{xx}(x^0(t),u^0(t),\lambda(t))x(t) \\ & + 2u(t)H_{ux}(x^0(t),u^0(t),\lambda(t))x^T(t) \\ & + u^T(t)H_{uu}(x^0(t),u^0(t),\lambda(t))u(t)\}dt \ge 0.\end{aligned} \tag{2.11}$$

If the attainability cone A specified in (2.1) satisfies $A = R^n \times R^{n_1}$ and condition (2.2) holds, then $\lambda_0 \neq 0$.

<u>Proof:</u> This is a special case of Theorem V.3.8. □

<u>Remark 2.2</u> A result due to Bernstein/Gilbert [1980, Theorem 4.3] shows

that for $\Omega = R^m$, $h \equiv 0$, the equality $A = R^n \times R^{n_1}$ holds iff the corresponding surjectivity condition for the steady state problem holds (cp. the discussion after Proposition VII.2.6) Thus in this case, the conditions for $\lambda_0 \neq 0$ in the periodic Problem 1.1 and the steady state Problem 1.2 coincide.

Remark 2.3 Sufficient optimality conditions are given in Maffezzoni [1974], Bittanti/Locatelli/Maffezzoni [1974], Speyer/Evans [1984].

Remark 2.4 Numerical work on optimal periodic control is rather scarce. See, however, Speyer/Evans [1984], Speyer/Dannemiller/Walker [1985]. Speyer/Evans [1984] propose an algorithm following periodic solutions of the Hamiltonian system

$$\begin{aligned} \dot{x}(t) &= H_y(x^0(t),u(x,y),y(t)) \\ \dot{y}(t) &= H_x(x^0(t),u(x,y),y(t)) \end{aligned} \tag{2.12}$$

as $H(x(t),u(t),y(t),y(t)) \equiv c$ varies; here $H(x,u,y) := g(x,u)+y^T f(x,u)$ (no state constraints!) and additional assumptions guarantee that u is given uniquely as a function $u = u(x,y)$ by the maximum condition.

The algorithm starts at an optimal steady state and stops when the "stopping condition" for optimality of the period is satisfied (cp. Theorem IV.2.2).

By a result due to Weinstein (cp. the survey Rabinowitz [1982]), there are at least n periodic solutions of (2.12) with $H(x(t),u(x(t),y(t)),y(t) \equiv c$, $|c|$ small enough, if for $\hat{H}(x,y) := H(x,u(x,y),y)$

$$\hat{H}(0,0) = 0, \quad \hat{H}_{xy}(x,y) = 0$$

and

$$\begin{pmatrix} \hat{H}_{xx} & \hat{H}_{xy} \\ \hat{H}_{yx} & \hat{H}_{yy} \end{pmatrix}$$

is positiv definite.

This, however, is only a *lower* bound on the number of periodic solutions.

Remark 2.5 The optimal periodic solution may be unstable and other pathological behaviour may occur under periodic forcing (cp. for chemical reactor studies e.g. Sincic/Bailey [1977], Matsubara/Onogi [1978a]). Hence the problem of stabilization around a periodic solution occurs (cp. Matsubara/Onogi [1978b]).

The linear-quadratic approach has been followed in Bittanti/Colaneri/Guarbadassi [1984] and Bittanti/Fronza/Guarbadassi [1972], Hewer [1975], Bittanti/Guarbadassi/Maffezoni/Silverman [1978], Kano/Nishimura [1979]; see also Brunovsky [1969]. For infinite dimensional problems see Da Prato [1987].

3. Local Properness under State Constraints

This section presents a Π-Test for local properness. Compared to Theorem VII.3.3, a state constraint is added and the required constraint qualification is discussed in detail.

Let (x^o,u^o) be locally optimal for Problem 1.2 and define the set Λ of steady state Lagrange multipliers by

$$\begin{aligned}\Lambda := \{0 \neq \lambda = (\lambda_o, y, y_1, z) \in R_+ \times R^n \times R^{n_1} \times R^\ell :\\ \lambda_o g_x(x^o,u^o) - y^T f_x(x^o,u^o) - y_1^T k_x(x^o,u^o) - z^T h_x(x^o) = 0\\ [\lambda_o g_u(x^o,u^o) - y^T f_u(x^o,u^o) - y_1^T k_u(x^o,u^o)][u-u^o] \geq 0\\ \text{for all } u \in \Omega\}.\end{aligned} \tag{3.1}$$

Recall that Problem 1.1 is a special case of Problem V.1.12 and that the Lagrangean L is given by (V.2.3). Here Lagrange multipliers λ have the form

$$\lambda = (\lambda_o, y^*, z^*) \in R \times R^{n+n_1} \times C(T;R^\ell)^*, \quad y^* = -(y, y_1) \tag{3.2}$$

and $\varphi = x(0) \in R^n$.

Observe that for Problem 1.1, condition (V.3.4) is equivalent to (2.4) - (2.8), and (V.3.5) - (V.3.7) are equivalent to the conditions on μ and $y(\cdot)$ stated in Theorem 2.1.

The crucial point for the proof of a Π-Test is to show that *all* Lagrange multipliers for Problem 1.1 come from Lagrange multipliers of Problem 1.2. This will certainly be the case if the Lagrange multipliers for Problem 1.1 are unique. We will indicate a condition, which ensures this property.

By Theorem II.1.19 Lagrange multipliers are unique if (II.1.6) and (II.1.7) hold. For Problem 1.1 condition (II.1.6) is equivalent to (2.2), and (II.1.7) is certainly satisfied if

$$\text{cl } H = C(T;R^{\ell}) \tag{3.3}$$

where $H := \{h_x(x^o)x(\cdot)$: there is $u \in U_{ad}(u^o) \cap [-U_{ad}(u^o)]$ such that

$$x(0) = x(\tau)$$
$$\dot{x}(t) = f_x(x^o,u^o)x(t) + f_u(x^o,u^o)u(t), \text{ a.a. } t \in T$$
$$0 = \int_0^\tau [k_x(x^o,u^o)x(t) + k_u(x^o,u^o)u(t)]dt\}.$$

<u>Theorem 3.1</u> Suppose that $(x^o,u^o) \in R^n \times R^m$ is a local optimal solution of Problem 1.2, assume that Hypothesis 1.3 and conditions (2.2), (3.3) hold and that

$$\text{rank } [jk\omega I - f_x(x^o,u^o)] = n \quad \text{for all} \quad k \in Z \tag{3.4}$$

where $\omega := 2\pi/\tau$.

Then (x^o,u^o) is locally proper if there exist

$$(x,u) \in C(0,\tau;R^n) \times L^\infty(0,\tau,R^m)$$

satisfying (2.4) - (2.8) such that

$$\max_{\lambda \in \Lambda} \int_0^\tau \{x^T(t)H_{xx}(x^o,u^o,\lambda)x(t) + 2u^T(t)H_{x,u}(x^o,u^o,\lambda)x(t) + u^T(t)H_{uu}(x^o,u^o,\lambda)u(t)\}dt < 0 \tag{3.5}$$

<u>Proof:</u> Follows as Theorem VII.3.1. □

Next we give a sufficient condition for condition (3.3) assuring uniqueness of Lagrange multipliers. For simplicity the analysis is restricted to the problem without isoperimetric and control constraints (i.e. $k \equiv 0$, $\Omega = R^m$).

Define

$$A := f_x(x^o,u^o), \quad B := f_u(x^o,u^o), \quad C := h_x(x^o). \tag{3.6}$$

<u>Lemma 3.2</u> Suppose that (3.4) holds and

$$\text{rank } C[jk\omega I - A]^{-1}B = \ell \quad \text{for all} \quad k \in Z. \tag{3.7}$$

Then (3.3) follows.

<u>Proof:</u> We may consider the complexified version of $C(T;R^{\ell})$. Then (3.3) is satisfied if for a base $a_1,\dots,a_\ell$ of C^ℓ,

$$a_i e^{jk\omega t} \in H.$$

This holds if

$$a_i \in \mathrm{range}[C(j\omega I-A)^{-1}B]$$

i.e. if (3.7) holds. □

Recall (3.6) and let

$$P := H_{xx}(x^0,u^0,\lambda), \quad Q := H_{xu}(x^0,u^0,\lambda), \quad R = H_{uu}(x^0,u^0,\lambda) \tag{3.8}$$

$$\Pi(\omega) = B^T(-j\omega I-A^T)^{-1}P(j\omega I-A)^{-1}B + Q^T(j\omega I-A)^{-1}B + B^T(-j\omega I-A)^{-1}Q + R. \tag{3.9}$$

We obtain the following Π-Test under state constraints.

Theorem 3.3 Suppose that $(x^0,u^0) \in R^n \times R^m$ is a local optimal solution of Problem 1.2, with $\Omega = R^m$ and $k = 0$. Let Hypothesis 1.3 and conditions (2.2), (3.4) and (3.7) be satisfied.
Then (x^0,u^0) is locally proper, if there exists $\nu \in R^m$ with $(\omega := 2\pi/\tau)$

$$\nu^T \Pi(\omega,\lambda)\nu < 0. \tag{3.10}$$

Proof: Follows from Theorem 3.1 and Lemma 3.2 (cp. Theorem VII.3.3). □

Remark 3.4 The Π-Test was proposed by Guarbadassi [1971], Bittanti/Fronza/Guarbadassi [1973] for unconstrained problems. A proof taking into account constraint qualifications was given by Bernstein/Gilbert [1980] (see also Chan/Ng [1979]) and extended to problems with control constraints by Bernstein [1985] (cp. also Guarbadassi/Schiavoni [1975]); the result by Bernstein was based on a generalization of Neustadt's theory of extremals in Bernstein [1984].

The Π-Test can be used to give a first estimate for the optimal periodic solution; see Sincic/Bailey [1980] for an application.

4. Example: Controlled Hopf Bifurcation in a Continuous Flow Stirred Tank Reactor (CSTR)

The purpose of the present section is to show that the scenario for local properness described in Chapter VIII actually occurs in "real" systems. We will analyse a simple example of a CSTR.

We beginn with a description of the model (following roughly Golubitsky/Schaeffer [1985]; cp. also Aris [1961], Douglas [1972]):

A reactant A flows at a constant rate r into a reactor vessel where a single exothermic reaction $A \to B$ takes place. We suppose that the reactor is well stirred, i.e. the concentration c_A of A and the temperature T are uniform throughout the vessel. The unused reactant and the product B leave the vessel at the same rate as the input, the concentration of the reactant and the temperature in the exit stream are equal to those in the reactor itself. The reactor is cooled by a coolant liquid of temperature T_c.

Using mass and energy balances, the concentration of A and the temperature in the reactor are modelled by the following pair of ordinary differential equations:

$$\frac{dT}{dt} = r(T_f-T) + k(T_c-T) + hZc_A\exp(-T_a/T), \tag{4.1}$$

$$\frac{dc_A}{dt} = r(c_f-c_A) - Zc_A\exp(-T_a/T); \tag{4.2}$$

here T_f and c_f are the temperature and concentration of the feed of reactant A, and k,h and Z are physical parameters, while T_f is the coolant temperature.

The factor $A(T) = \exp(-T_a/T)$, with $T_a = E/R$ (E the "activation energy", converted to a temperature by means of the universal gas constant R), is of Arrhenius form and governs the temperature dependence of the reaction rate.

We will consider k which is proportional to the overall heat transfer coefficient as a control variable which can be adjusted by the coolant flow rate (cp. Onogi/Matsubara [1980]).

An exhaustive study of the steady states of this equation requires the use of singularity theory (Golubitsky/Keyfitz [1980], Golubitsky/Schaeffer [1985]). This is beyond the scope of these notes. Instead we will content ourselves here with a study where, following Poore [1974], the Damkoehler number Z is taken as a bifurcation parameter and we will consider only a subset (region III in Poore's classification) of the possible parameter configurations.

The concentration c_B of the product B satisfies:

$$C_A + C_B = \bar{c} = \text{constant}.$$

Thus

$$\frac{d}{dt} C_B = - \frac{d}{dt} C_A = -r(c_f-c_A)hZc_A\exp(-T_a/T)$$

i.e.

$$\frac{d}{dt} c_B = -r(c_f+c_B-\bar{c}) + hZ(1-c_B)\exp(-T_a/T). \tag{4.3}$$

We convert (4.1), (4.3) to dimensionless form by defining

$$\gamma := T_a/T_f, \quad x_c = \gamma(T_f-T_c)/T_f \tag{4.4}$$

$$x_1 := \gamma(T-T_f)/T_f, \quad x_2 := (c_B+c_f-\bar{c})/c_f = (c_f-c_A)/c_f . \tag{4.5}$$

Then

$$\exp(-T_a/T) = \exp(-\gamma)\exp\frac{x_1}{1+x_1/\gamma} \tag{4.6}$$

and (4.1), (4.3) become

$$\frac{dx_1}{dt} = -rx_1 + k[x_c-x_1] + hZ\exp(-\gamma)/T_f[1-x_2]\exp\frac{x_1}{1+x_1/\gamma} \tag{4.7}$$

$$\frac{dx_2}{dt} = -rx_2 + Z\exp(-\gamma)/T_f[1-x_2]\exp\frac{x_1}{1+x_1/\gamma} . \tag{4.8}$$

Under the transformation $t \to rt$, one can rewrite (4.7),(4.8) as

$$\frac{dx_1}{dt} = -x_1 + k/r[x_c-x_1] + hZ\exp(-\gamma)/(rT_f)[1-x_2]\exp\frac{x_1}{1+x_1/\gamma} \tag{4.9}$$

$$\frac{dx_2}{dt} = -x_2 + Z\exp(-\gamma)/(rT_f)[1-x_2]\exp\frac{x_1}{1+x_1/\gamma} . \tag{4.10}$$

For large activation energy one may consider the limiting case $\gamma \to \infty$ (cp. Poore [1974], Uppal/Ray/Poore [1974]), so that the nonlinearity takes the form $\exp(x_1)$.

We consider the following optimal periodic control problem.

Problem 4.1 Minimize $-1/\tau \int_0^\tau x_2(s)ds$

$$\text{s.t.} \quad \dot{x}_1(t) = -x_1(t)-u(t)[x_1(t)-x_c] + B\alpha[1-x_2(t)]\exp(x_1(t))$$

$$\dot{x}_2(t) = -x_2(t) + \alpha[1-x_2(t)]\exp(x_1(t)), \qquad \text{a.a. } t \in [0,\tau]$$

$$x_1(0) = x_1(\tau), \quad x_2(0) = x_2(\tau)$$

$$1/\tau \int_0^\tau u(s)ds = \beta$$

with $\alpha = Z\exp(-\gamma)/(rT_f)$, $B = h$, $u = k/r$.

Thus we want to maximize the average product concentration, while keeping the average heat transfer coefficient constant.

The corresponding optimal steady state problem has the following form.

Problem 4.2 Minimize $-x_2$

$$\text{s.t.}\qquad 0 = -x_1 - u(x_1-x_c) + B\alpha(1-x_2)e^{x_1}$$

$$0 = -x_2 + \alpha(1-x_2)e^{x_1}$$

$$0 = u - \beta$$

with α, β, B and x_c as above.

Thus here one has to minimize only over x_1, x_2, and $u = \beta$.

First we determine the equations for the steady states of the system. Starting from

$$0 = -x_1^s - \beta(x_1^s - x_c) + B\alpha(1-x_2^s)e^{x_1^s}$$

$$0 = -x_2^s \qquad\qquad + \alpha(1-x_2^s)e^{x_1^s}$$

we obtain

$$0 = x_1^s + \beta(x_1^s - x_c) - Bx_2^s = (1+\beta)x_1^s - \beta x_c - Bx_2^s$$

and hence

$$x_1^s = \frac{B}{1+\beta}x_2^s + \frac{\beta x_c}{1+\beta} \tag{4.11}$$

$$\alpha = \frac{x_2^s}{1-x_2^s}e^{-x_1^s} = \frac{x_2^s}{1-x_2^s}\exp\{-\frac{B}{1+\beta}x_2^s - \frac{\beta x_c}{1+\beta}\} \tag{4.12}$$

Thus (x_1^s, x_2^s) is a steady state of the system equation in Problem 4.1 or, equivalently, satisfies the constraints of Problem 4.2, if it satisfies (4.11) and (4.12). Note that x_2^s determines x_1^s uniquely via (4.11).

The following classification follows by an elementary analysis of (4.12) (Poore [1974, Theorem 3.1])

Proposition 4.3 Let

$$m^{\pm} := 1/2 \pm 1/2\sqrt{1 - 4(1+\beta)/B} \tag{4.13}$$

$$\alpha^{\pm} := m^{\pm}/(1-m^{\pm})\exp\{-Bm^{\pm}/(1+\beta) - \beta x_c/(1+\beta)\} \tag{4.14}$$

Then the following assertions hold:

(i) If $B \le 4(1+\beta)$ or if $B > 4(1+\beta)$ and $\alpha \in (0,\alpha^-)\cup(\alpha^+,\infty)$, then there exists a unique solution (x_1,x_2) of (4.11),(4.12).

(ii) If $B > 4(1+\beta)$ and $\alpha = \alpha^+$ or $\alpha = \alpha^-$, there exist exactly two solutions (x_1,x_2) of (4.11),(4.12).

(iii) If $B > 4(1+\beta)$ and $\alpha \in (\alpha^-,\alpha^+)$ there exist exactly three solutions of (4.11),(4.12).

Fig. 1 illustrates a typical situation for $B > 4(1+\beta)$

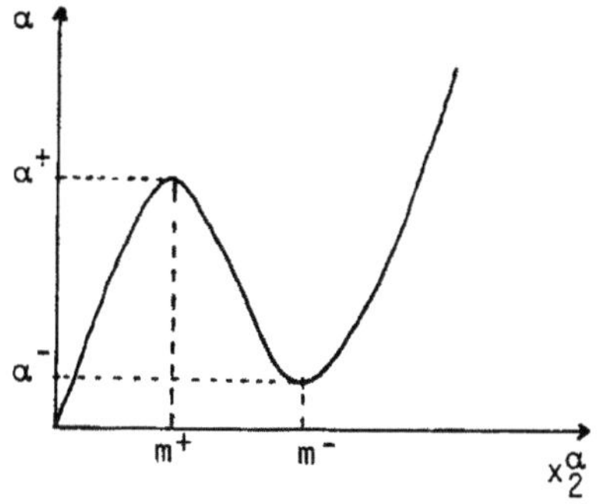

Fig. 1

Linearizing the system equation around (x_1^s, x_2^s, u^s), $u^s = \beta$ one gets

$$\begin{aligned}\dot{x}_1(t) &= -(1+\beta)x_1(t) + B\alpha(1-x_2^s)e^{x_1^s}x_1(t) - B\alpha e^{x_1^s}x_2(t) - [x_1^s - x_c]u(t)\\ \dot{x}_2(t) &= -x_2(t) + \alpha(1-x_2^s)e^{x_1^s}x_1(t) - \alpha e^{x_1^s}x_2(t),\end{aligned} \tag{4.15}$$

i.e.

$$\begin{pmatrix}\dot{x}_1(t)\\ \dot{x}_2(t)\end{pmatrix} = \begin{pmatrix}-(1+\beta)+B\alpha(1-x_2^s)e^{x_1^s} & -B\alpha e^{x_1^s}\\ \alpha(1-x_2^s)e^{x_1^s} & -1-\alpha e^{x_1^s}\end{pmatrix}\begin{pmatrix}x_1(t)\\ x_2(t)\end{pmatrix} + \begin{pmatrix}x_c - x_1^s\\ 0\end{pmatrix}u(t).$$

Defining

$$A := \begin{bmatrix} Bx_2^s - 1 - \beta & -B\dfrac{x_2^s}{1-x_2^s} \\ x_2^s & -\dfrac{1}{1-x_2^s} \end{bmatrix}, \quad B_0 := \begin{bmatrix} x_c - x_1^s \\ 0 \end{bmatrix}, \quad x := \begin{bmatrix} x_1 \\ x_2 \end{bmatrix} \tag{4.16}$$

and using (4.12) one can write (4.15) as

$$\dot{x}(t) = Ax(t) + B_0 u(t). \tag{4.17}$$

For any 2×2 matrix $A = (a_{ij})$, the eigenvalues $\lambda_{1,2}$ are given by

$$\lambda_{1,2} = 1/2\mathrm{tr}A \pm 1/2\sqrt{(\mathrm{tr}A)^2 - 4\det A}. \tag{4.18}$$

Thus the eigenvalues of A are on the imaginary axis and unequal zero iff

$$\mathrm{tr}A = 0 \quad \text{and} \quad \det A > 0.$$

Then

$$\lambda_{1,2} = \pm j\omega_0 \quad \text{where} \quad \omega_0 = \sqrt{\det A}.$$

Now $\mathrm{tr}A = 0$ means

$$Bx_2^s - 1 - \beta - \frac{1}{1-x_2^s} = 0$$

or, equivalently, $x_2^s \neq 1$ and

$$Bx_2^s - B(x_2^s)^2 - 1 + x_2^s - \beta + \beta x_2^s - 1 = 0$$

i.e.

$$B(x_2^s)^2 - (B+1+\beta)x_2^s + 2 + \beta = 0 \tag{4.19}$$

and hence, if real, the roots

$$s^{\pm} = \frac{1}{2B}(B+1+\beta) \pm \frac{1}{2B}\sqrt{(B+1+\beta)^2 - 4B(2+\beta)} \tag{4.20}$$

determine those steady states $(x_1^s, x_2^s) = (\frac{B}{1+\beta}s^{\pm} + \frac{\beta x_c}{1+\beta}, s^{\pm})$ of the system equation, for which the linearized system satisfies $\mathrm{tr}A = 0$.

We restrict our attention to a subset of the parameter configurations.

<u>Proposition 4.4</u> Let

$$B > (1+\beta)^3/\beta, \quad \text{and} \quad 1 \leq \beta < \infty. \tag{4.21}$$

Then for $\alpha \in (\alpha^-,\alpha^+)$ there exist exactly three solutions of (4.11), (4.12) and for $\alpha \in (0,\alpha^-)\cup(\alpha^+,\infty)$ there exists a unique solution of (4.11),(4.12). Furthermore

$$0 < m^- < s^- < m^+ < s^+ < 1. \tag{4.22}$$

The critical point is an asymptotically stable node or spiral for $x_2 \in (0,m^-)\cup(s^+,1)$, an unstable saddle for $x_2 \in (m_1,m_2)$, and an unstable node or saddle for $x_2 \in (m^+,s^+)$.

Proof: The multiplicity assertion follows from Proposition 4.3, noting that (4.19) implies

$$B > (1+\beta)(1+\beta)^2/\beta \geq 4(1+\beta).$$

The inequalities (4.22) follow by inspection of (4.13) and (4.20), and the last assertion follows by Poore [1974,Theorem 4.1(iii)].

□

As a consequence of Proposition 4.4 we obtain that, if (4.21) holds, $x_2^s = s^+$ (globally) maximizes x_2 in Problem 4.2 (which anyway has only three admissible points).

Proposition 4.5 Consider the linearized system matrix A at the steady state

$$x_1^s = B/(1+\beta)s^+ + \beta x_c/(1+\beta), \quad x_2^s = s^+. \tag{4.23}$$

Then $\det A > 0$, if (4.21) holds.

Proof: Note that $m^{\pm}$ are the roots of

$$\det A = (B(x_2^s)^2 - Bx_2^s + 1+\beta)/(1-x_2^s) = 0.$$

Hence the assertion follows, since $1 > s^+ > m^+$.

□

Thus we know that at the steady state defined by (4.23) the matrix A of the linearized system has roots $\pm j\omega_o$, $\omega_o \neq 0$.

We show next, that the rank condition (VIII.2.5) holds.

Note that in a neighbourhood of $x_2^s = s^+$, equation (4.12) determines x_2^s uniquely as a function of α; hence we write x_2^α for this solution of (4.12) and x_1^α for the corresponding solution of (4.11); similarly

$$A^{\alpha} := \begin{bmatrix} Bx_2^{\alpha} - 1 - \beta & -B \dfrac{x_2^{\alpha}}{1-x_2^{\alpha}} \\ x_2^{\alpha} & -\dfrac{1}{1-x_2^{\alpha}} \end{bmatrix} . \tag{4.24}$$

Let α^0 be the parameter value corresponding to $x_2^{\alpha^0} = s^+$, i.e.

$$\alpha^0 = \frac{x_2^{\alpha^0}}{1-x_2^{\alpha^0}} e^{-x_1^{\alpha^0}} = \frac{s^+}{1-s^+} \exp\{-\frac{B}{1+\beta} s^+ - \frac{\beta}{1+\beta} x_c\} . \tag{4.25}$$

For $|\alpha-\alpha^0|$ small, $x_1^{\alpha}, x_2^{\alpha}$, A^{α}, $\mathrm{tr} A^{\alpha}$ and $\det A^{\alpha}$ depend continuously on α.

Hence (4.21) implies that for $|\alpha-\alpha^0|$ small enough that

$$(\mathrm{tr} A^{\alpha})^2 - 4 \det A^{\alpha} < 0$$

and the eigenvalues $\lambda_{1,2}(\alpha)$ of A^{α} satisfy

$$2 \operatorname{Re} \lambda_1(\alpha) = 2 \operatorname{Re} \lambda_2(\alpha) = \mathrm{tr}\, A^{\alpha} = Bx_2^{\alpha} - 1 - \beta - \frac{1}{1-x_2^{\alpha}} . \tag{4.26}$$

For (VIII.2.5) it suffices that

$$\frac{d}{d\alpha} \operatorname{Re} \lambda_1(\alpha)\Big|_{\alpha=\alpha^0} \neq 0. \tag{4.27}$$

By implicit differentiation of (4.12), one gets using (4.11) and (4.12) again,

$$\begin{aligned} 1 &= [B(x_2^{\alpha})^2 - Bx_2^{\alpha} + 1+\beta]/(1-x_2^{\alpha}) \frac{d}{d\alpha} x_2^{\alpha}\, \alpha/x_2^{\alpha}\, 1/(1+\beta) \\ &= \det A^{\alpha}\, \alpha/x_2^{\alpha}\, 1/(1+\beta) \frac{d}{d\alpha} x_2^{\alpha}. \end{aligned} \tag{4.28}$$

Thus by Proposition 4.5

$$\frac{dx_2^{\alpha}}{d\alpha}\Big|_{\alpha=\alpha^0} > 0.$$

Differentiation of (4.26) yields

$$\begin{aligned} \frac{d}{d\alpha} 2 \operatorname{Re} \lambda_1(\alpha)\Big|_{\alpha=\alpha^0} &= [B-1/(1-s^+)^2] \frac{d}{d\alpha} x_2^{\alpha} \\ &= [B(s^+)^2 - 2Bs^+ + B-1]/(1-s^+)^2 \frac{d}{d\alpha} x_2^{\alpha} ; \end{aligned}$$

but the first factor is by (4.21) less than

$$B(s^+)^2 - (B+1)s^+ + B-1 = [B(s^+)^2 - (B+1+\beta)s^+ + 2+\beta] + \beta s^+ - (1+\beta) < -1.$$

Hence $\frac{d}{d\alpha} \operatorname{Re} \lambda_1(\alpha)|_{\alpha=\alpha^0} < 0$ and condition (VIII.2.5) is satisfied.

Next we check that the controllability condition (VIII.2.7) holds, i.e.

$$\operatorname{Adj}[j\omega_0 I - A^{\alpha^0}] \begin{pmatrix} x_c - s^+ \\ 0 \end{pmatrix} \neq 0 \tag{4.29}$$

We abbreviate

$$a := s^+ = x_2^{\alpha^0}, \quad b := Bs^+ - 1 - \beta = Ba - 1 - \beta. \tag{4.30}$$

Observe

$$b = 1/(1-a), \quad ab = a/(1-a) \tag{4.31}$$

$$b(1-a) = 1, \quad ab+1 = b$$

$$Bab = Bs^+/(1-s^+).$$

$$A^{\alpha_0} = \begin{pmatrix} b & -Bab \\ a & -b \end{pmatrix} \tag{4.32}$$

$$0 < \omega_0^2 = \det A^{\alpha^0} = Ba^2b - b^2 = b(Ba^2 - b). \tag{4.33}$$

We find

$$\operatorname{Adj}[j\omega_0 - A^{\alpha^0}] \begin{pmatrix} x_c - s^+ \\ 0 \end{pmatrix} = \begin{pmatrix} j\omega_0 - b & Bab \\ -a & -j\omega_0 + b \end{pmatrix} \begin{pmatrix} x_c - s^+ \\ 0 \end{pmatrix}$$

$$= (x_c - s^+) \begin{pmatrix} j\omega_0 - b \\ -a \end{pmatrix} \neq 0, \quad \text{if} \quad s^+ \neq x_c.$$

The required constraint qualification, Hypothesis VIII.1.5, holds since

$$\det \left(\begin{array}{cc|c} \multicolumn{2}{c|}{A^{\alpha^0}} & B^{\alpha^0} \\ \hline 0 & 0 & 1 \end{array} \right) = \det A(\alpha^0) \neq 0.$$

It remains to discuss the local properness condition. Note that the matrix A^{α^0} has eigenvectors $p_1' = (p_{11}, p_{12})^T$ determined by

$$-ap_{11} + (j\omega_0 + b)p_{12} = 0$$

i.e.

$$p_1 = \begin{pmatrix} j\omega_0 + b \\ 0 \end{pmatrix} \tag{4.34}$$

is an eigenvector.

The function H has the following form (suppressing the argument λ_0, since $\lambda_0 = 1$):

$$\begin{aligned} H(x_1, x_2, u, \lambda_1, \lambda_2, \lambda_3) = -x_2 &+ \lambda_1[-x_1 - u(x_1 - x_c) + B\alpha(1-x_2)e^{x_1}] \\ &+ \lambda_2[-x_2 + \alpha(1-x_2)e^{x_1}] \\ &+ \lambda_3[u - \beta] \ . \end{aligned} \tag{4.35}$$

For the derivatives at (x_1, x_2, β) we get

$$H_{x_1} = -(1+\beta)\lambda_1 + [\lambda_1 B + \lambda_2]\alpha(1-x_2)e^{x_1} \tag{4.36}$$

$$H_{x_2} = -1 - \lambda_2 - [\lambda_1 B + \lambda_2]\alpha e^{x_1}$$

$$H_u = -\lambda_1 x_1 + \lambda_3 \ .$$

Thus by (4.12)

$$0 = H_{x_1} = -(1+\beta)\lambda_1 + [\lambda_1 B + \lambda_2]a \tag{4.37}$$

$$0 = H_{x_2} = -1 - \lambda_2 - [\lambda_1 B + \lambda_2]a/(1-a) \tag{4.38}$$

$$\lambda_3 = \lambda_1 x_1^{\alpha^0} \ . \tag{4.39}$$

We compute further from (4.36)

$$H_{x_1 x_1} = [\lambda_1 B + \lambda_2]\alpha(1-x_2)e^{x_1}$$

$$H_{x_1 x_2} = -[\lambda_1 B + \lambda_2]\alpha e^{x_1} = H_{x_2 x_1}$$

$$H_{x_2 x_2} = 0 \ .$$

Thus

$$H_{xx} = \begin{bmatrix} H_{x_1x_1} & H_{x_1x_2} \\ H_{x_2x_1} & H_{x_2x_2} \end{bmatrix} = [\lambda_1 B+\lambda_2]\alpha e^{x_1} \begin{bmatrix} 1-x_2 & -1 \\ -1 & 0 \end{bmatrix} .$$

This, evaluated at $x_1^{\alpha_o}, x_2^{\alpha_o}$ gives

$$H_{xx} = [\lambda_1 B+\lambda_2]\alpha e^{x_1^{\alpha_o}} \begin{pmatrix} 1-a & -1 \\ -1 & 0 \end{pmatrix} \tag{4.40}$$

Next we compute the Lagrange multipliers from (4.37) and (4.38). By (4.38)

$$0 = -(1+\lambda_2)(1-a) - (\lambda_1 B+\lambda_2)a.$$

Adding (4.37)

$$0 = -(1+\beta)\lambda_1 - (1+\lambda_2)(1-a) = (1+\beta)\lambda_1 + 1-a + \lambda_2 - a\lambda_2.$$

Thus

$$\lambda_2 = -[(1+\beta)\lambda_1 + 1-a]/(1-a) = -1-(1+\beta)\lambda_1/(1-a) \tag{4.41}$$

and insertion into (4.37) gives

$$0 = -(1+\beta)\lambda_1 + [\lambda_1 B - ((1+\beta)\lambda_1 + 1-a)/(1-a)]a$$

i.e.

$$\begin{aligned} 0 &= (1+\beta)\lambda_1 - \lambda_1 Ba + [(1+\beta)\lambda_1 + 1-a]a/(1-a) \\ &= [1+\beta - Ba + ab(1+\beta)]\lambda_1 + a \\ &= [Ba-b-Ba \quad + ab(Ba-b)]\lambda_1 + a \\ &= b[Ba^2-ab \quad - 1]\lambda_1 + a \\ &= b[Ba^2-b]\lambda_1 + a \end{aligned}$$

by using (4.30) and repeatedly (4.31).

Thus

$$\lambda_1 = -\frac{a}{b(Ba^2-b)} . \tag{4.42}$$

Now (4.41) and (4.42) yield

$$\begin{aligned} \lambda_1 B + \lambda_2 = \lambda_1 B - 1-(1+\beta)\lambda_1/(1-a) &= -1 + \lambda_1[B-(Ba-b)b] \\ &= -1 - \frac{a(B-Bab+b^2)}{b(Ba^2-b)} . \end{aligned} \tag{4.43}$$

On the other hand one computes

$$\bar{p}_1^T\, H_{xx}\, p_1\; 1/\alpha\; e^{-x_1^{\alpha^0}} / [\lambda_1 B+\lambda_2]$$

$$= [-j\omega_0+b \quad a] \begin{pmatrix} 1-a & -1 \\ -1 & 0 \end{pmatrix} \begin{pmatrix} j\omega_0+b \\ a \end{pmatrix}$$

$$= [(-j\omega_0+b)(1-a)-a \quad -(-j\omega+b)] \begin{pmatrix} j\omega_0+b \\ a \end{pmatrix}$$

$$= (-j\omega_0+b)(1-a)(j\omega_0+b)-a(j\omega_0+b) - (-j\omega_0+b)a$$

$$= (\omega_0^2+b^2)(1-a) - 2ab$$

$$= Ba^2b(1-a) - 2ab \qquad \text{by (4.33)}$$

$$= a(Ba-2b). \qquad \text{by (4.30)}$$

Thus

$$\bar{p}_1^T\, H_{xx}\, p_1 < 0 \tag{4.44}$$

iff

$$Ba - 2b > 0 \quad \text{and} \quad \lambda_1 B + \lambda_2 < 0 \tag{4.45}$$

or

$$Ba - 2b < 0 \quad \text{and} \quad \lambda_1 B + \lambda_2 > 0. \tag{4.46}$$

We deal only with the case (4.45).
Suppose that

$$Ba - 2b > 0. \tag{4.47}$$

Then

$$\begin{aligned} a(B - Bab + b^2) &> 2b - Ba^2b + ab^2 \\ &= b(2 - Ba^2 + ab) \\ &= b(1 + ab - Ba^2 + 1) \\ &= b(b - Ba^2 + 1) \qquad \text{by (4.31).} \end{aligned}$$

Hence by (4.43)

$$\lambda_1 B + \lambda_2 = -1 - \frac{a(B-Bab+b^2}{b(Ba^2-b)} < 1 - \frac{b(b-Ba^2+1)}{b(Ba^2-b)}$$

$$= -1 + 1 - 1/(Ba^2-b) < 0.$$

Thus (4.47) implies (4.45) and hence the local properness condition (4.44).

(We would have been glad to prove that

$$Ba - 2b < 0$$

implies (4.46). However, arguing along the same lines as above, one can only show that $Ba - b < 0$ implies $\lambda_1 B + \lambda_2 > 0$; observe that $Ba - b < 0$ is never satisfied.)

Next we supply sufficient conditions in terms of the system parameters β and B for (4.47).

Using (4.31), condition (4.47) can be reformulated as

$$Ba > 2/(1-a)$$

or

$$Ba^2 - Ba + 2 < 0. \tag{4.48}$$

But $a = s^+$ is a solution of (4.19), that is

$$\begin{aligned} 0 &= Ba^2 - (B+1+\beta)a + 2 + \beta \\ &= Ba^2 - Ba + 2 + \beta - (1+\beta)a. \end{aligned}$$

Thus (4.48) is equivalent to

$$\beta - (1+\beta)a > 0$$

or

$$s^+ < \beta/(1+\beta). \tag{4.49}$$

By (4.20), this is equivalent to

$$\frac{1}{2B}(B+1+\beta)^2 - 4B(2+\beta) < \beta/(1+\beta) - \frac{1}{2B}(B+1+\beta). \tag{4.50}$$

The right hand side is nonnegative iff

$$\begin{aligned} \beta/(1+\beta) - \tfrac{1}{2B}(B+1+\beta) &= 1/[2B(1+\beta)][2B\beta - B(1+\beta) - (1+\beta)^2] \\ &= 1/[2B(1+\beta)][B(\beta-1) - (1+\beta)^2] > 0 \end{aligned}$$

thus inequality (4.50) is satisfied iff

$$B > (1+\beta)^2/(\beta-1), \quad \beta > 1 \tag{4.51}$$

and

$$\frac{1}{4B^2}[(B+1+\beta)^2 - 4B(2+\beta)] < \frac{1}{4B^2(1+\beta)^2}[B\beta-B-(1+\beta)^2]^2 \tag{4.52}$$

(4.52) is equivalent to

$$
\begin{aligned}
&(1+\beta)^2[B^2+2B(1+\beta)+(1+\beta)^2 - 4B(2+\beta)] \\
&= (1+\beta)^2B^2 + 2B(1+\beta)^3 + (1+\beta)^4 - 4B(2+\beta)(1+\beta)^2 \\
&< B^2\beta^2 - 2B^2\beta - 2B\beta(1+\beta)^2 + B^2 + 2B(1+\beta)^2 + (1+\beta)^4 \\
&= (1+\beta^2)B^2 - 2B^2\beta + (2B-2B\beta)(1+\beta)^2 + (1+\beta)^4
\end{aligned}
$$

i.e.

$$
\begin{aligned}
0 &> 2B(1+\beta)^3 - (8B+4B\beta)(1+\beta)^2 + 2B^2\beta + (2B\beta-2B)(1+\beta)^2 \qquad (4.53) \\
&= B[2(1+\beta)^3 - 2(\beta+5)(1+\beta)^2 + 2B\beta].
\end{aligned}
$$

Condition (4.21) implies

$$
\begin{aligned}
2(1+\beta)^3 - 2(\beta+5)(1+\beta)^2 + 2B\beta &< 4B\beta - 2(\beta+5)(1+\beta)^2 \\
&= 2[2B\beta - (\beta+5)(1+\beta)]^2.
\end{aligned}
$$

Thus assuming (4.21), inequality (4.53) holds iff

$$B < (5+\beta)(1+\beta)^2/2\beta \qquad (4.54)$$

is satisfied.

Resuming the results above, we arrive at the following theorem.

<u>Theorem 4.6</u> Let $\beta > 1$ and assume

$$\max\{(1+\beta)^3/\beta, (1+\beta)^2/(\beta-1)\} < B < (5+\beta)(1+\beta)^2/2\beta. \qquad (4.55)$$

Let s^+ and α^0 be defined by (4.20) and (4.25), respectively and assume $s^+ \neq x_c$. Then there exist neighborhoods of α^0 and of $(x_1^{\alpha^0}, x_2^{\alpha^0}) = (B/(1+\beta)s^+ + \beta x_c/(1+\beta), s^+)$ such that in these neighborhoods the steady state solutions $x = (x_1^\alpha, x_2^\alpha)$ of Problem 4.2 are unique, depend in a continuously differentiable way on α, and $(x^\alpha, u^\alpha) = (x^\alpha, \beta)$ are locally proper.

Some tedious, but straightforward computations allow to determine $\Pi(\omega,\alpha)$ in this example.

One obtains (for $x_c = 0$)

$$\Pi(\omega,\alpha) = \left\{ \frac{B^2}{(1+\beta)^2} (x_2^\alpha)^3(1-x_2^\alpha)\left[\frac{1-2x_2^\alpha}{1-x_2^\alpha} + \omega^2(1-x_2^\alpha)\right] \frac{b^2+ab^2+aB-2a^2+B}{bBa^2-b^2} \right.$$

$$- \frac{2x_2^\alpha aB}{(1+\beta)(bBa^2-b^2)} \left[\omega^2(1-x_2^\alpha) + Bx_2^\alpha - 1-\beta - B(x_2^\alpha)^2\right]\Bigg\}$$

$$\Bigg/ \Big\{[\omega^2(1-x_2^\alpha) + Bx_2^\alpha - 1-\beta - B(x_2^\alpha)^2]^2 + [1-(Bx_2^\alpha-1-\beta)(1-x_2^\alpha)]^2\Big\} .$$

For the parameter values $B = 15$, $\beta = 2$ a controlled Hopf bifurcation occurs at $\alpha = \alpha^0 \approx 0.1356$. Figure 2 shows $\Pi(\omega,\alpha)$ for $0 < \omega < 15$, $0.06 < \alpha < 0.25$ (I thank Bernd Kelb and Joachim Schalthöfer for producing this plot). As for the retarded Liênard equation in Section VIII.3, a pole occurs at $\alpha = \alpha_0$, $\omega = \omega_0$ (here a linear approximation in (4.12) has been used in order to compute x_2^α for α near α^0).

Remark 4.7 Both direction of the Hopf bifurcation occur in the region specified by (4.55) (cp. Poore [1974]).

Remark 4.8 Refinements of the model equation considered here include delays in the state variables (cp. Ray/Soliman [1971], Soliman/Ray [1972a,b]). It would be interesting to extend the discussion above to such a model.

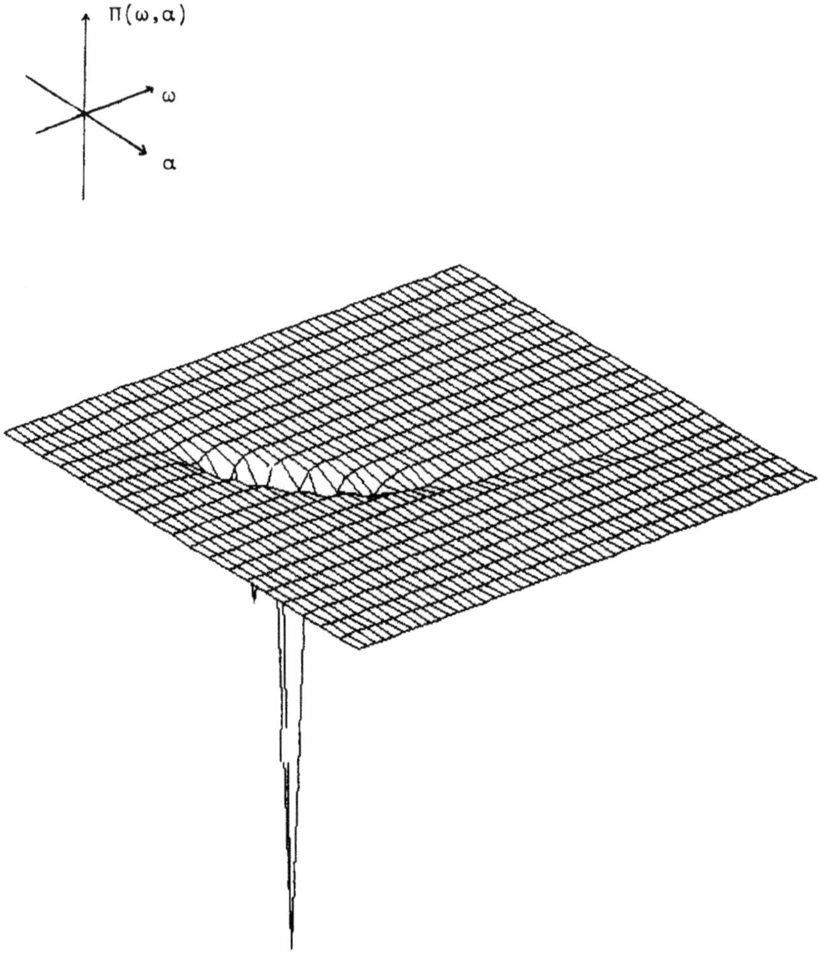

Fig. 2 shows $\Pi(\omega,\alpha)$ for $0 < \omega < 15$, $0.06 < \alpha < 0.25$

REFERENCES

Alt, W. [1979], Stabilität mengenwertiger Abbildungen mit Anwendungen auf nichtlineare Optimierungsprobleme, Dissertation, Bayreuther Mathematische Schriften, Heft 3.

Alt, W. [1983], Lipschitzean perturbations of infinite optimization problems, Mathematical Programming with Data Perturbations II, A.V. Fiacco ed., Marcel Dekker, New York, 7-21.

Aris, R. [1961], The Optimal Design of Chemical Reactions, Academic Press, New York and London.

Aris, R., Nemhauser, G.L., Wilde, D.J. [1964], Optimization of multistage cyclic and branching systems by serial procedures, A.I.Ch.E. Journal, 10, 913-919.

Aubin, J.P. Ekeland, I. [1984], Applied Nonlinear Analysis, J. Wiley, New York.

Bader, G. [1983], Numerische Behandlung von Randwertproblemen für Funktionaldifferentialgleichungen, Institut für Angewandte Mathematik, Universität Heidelberg, Preprint, 227.

Bailey, J.E. [1973], Periodic operation of chemical reactors: A review, Chem. Eng. Commun., 1, 111-124.

Bank, B., Guddat, J., Klatte, D., Kummer, B., Tammer, K. [1982], Nonlinear Parametric Optimization, Akademie-Verlag, Berlin.

Banks, H.T. [1972], Control of functional differential equations with function space boundary conditions, Delay and Functional Differential Equations and their Applications, K. Schmitt (ed.), Academic Press, 1-16.

Banks, H.T., Jacobs, M.Q. [1973], An attainable sets approach to optimal control of functional differential equations with function space boundary conditions, J. Diff. Equations, 13, 127-149.

Banks, H.T., Jacobs, M.Q., Langenhop, C.E. [1974], Function space controllability for linear functional differential equations, Differential Games and Control Theory, E.O. Roxin, P. Liu, R.L. Sternberg, eds., Dekker, 81-98.

Banks, H.T., Jacobs, M.Q., Langenhop, C.E. [1975], Characterization of the controlled states in W_2^1 of linear hereditary systems, SIAM J. Control, 13, 611-649.

Banks, H.T., Kent, G.A. [1972], Control of functional differential equations of retarded and neutral type to target sets in function space, SIAM J. Control, 10, 567-593.

Banks, H.T., Manitius, A. [1974], Application of abstract variational theory to hereditary systems - a survey, IEEE Trans. Aut. Control, 524-533.

Barbanti, L. [1980], Liênard equations and control, Functional Differential Equations and Bifurcation, A.F. Izê, ed., Lecture Notes in Math., Vol. 199, Springer-Verlag, 1-22.

Barbu, V., Precupanu, Th. [1978], Convexity and Optimization in Banach Spaces, Sijthoff and Noordhoff, Alphen aan de Rijn.

Bartosiewicz, Z. [1979], Closedness of the attainable set of the linear neutral control system, Control Cybernet., 8.

Bartosiewicz, Z. [1984], A criterion of closedness of an attainable set of a delay system, Systems Control Lett., 3, 211-215.

Bartosiewicz, Z., Sienkiewicz, G. [1984], On closedness of the attainable set of a delay system, Systems Control Lett., 4, 293-300.

Bates, G.R. [1977], Herditary Optimal Control Problems, Ph.D. Thesis, Purdue University, West Lafayette, Ind.

Ben-Tal, A., Zowe, J. [1982], A unified theory of first and second order conditions for extremum problems in topological vector spaces, Math. Programming Study, No. 19, 39-76.

Berger, M.S. [1977], Nonlinearity and Functional Analysis, Academic Press.

Berkovitz, L.D. [1974], Optimal Control Theory, Springer Verlag.

Berkovitz, L.D. [1975], A penalty function proof of the maximum principle, Appl. Math. Optim. 2, 291-303.

Bernier, C., Manitius, A. [1978], On semigroups in $R^n \times L^p$ corresponding to differential equations with delay, Can. J. Math. 30, 897-914.

Bernstein, D.S. [1984], A systematic approach to higher order necessary conditions in optimization theory, SIAM J. Control Optim. 22, 211-238.

Bernstein, D.S. [1985], Control constraints, abnormality and improved performance by periodic control, IEEE Trans. Aut. Control, AC-30, 367-378.

Bernstein, D.S., Gilbert, E.G. [1980], Optimal periodic control: The Π Test revisited, IEEE Trans. Aut. Control, AC-25, 673-684.

Bien, Z. [1975], Optimal control of delay systems, Ph.D. Thesis, University of Iowa, Ames, Io.

Bien, Z., Chyung, D.H. [1980], Optimal control of delay systems with a final function condition, Internat. J. Control, 32, 539-560.

Bittanti, S., Colaneri, P., Guarbadassi, G. [1984], H-controllability and observability of linear periodic systems, SIAM J. Control Optim. 22, 889-901.

Bittanti, S., Fronza, G., Guarbadassi, G. [1972], Periodic optimization of linear systems, in Marzollo (ed.) [1972], 213-231.

Bittanti, S., Fronza, G., Guarbadassi, G. [1973], Periodic control: A frequency domain approach, IEEE Trans. Aut. Control, AC-18, 33-38.

Bittanti, S., Fronza, G., Guarbadassi, G. [1974], Discrete periodic optimization, Automation and Remote Control, 35, No. 10.

Bittanti, S., Fronza, G., Guarbadassi, G. [1976], Optimal steady state versus periodic operation in discrete systems, J. Opt. Theory Appl. 521-536.

Bittanti, S., Guarbadassi, G., Maffezzoni, C., Silverman, C. [1978], Periodic systems: controllability and the matrix Riccati equation, SIAM J. Control Optim. 16, 37-40.

Bittanti, S., Locatelli, A., Maffezzoni, C. [1974], Second variation methods in periodic optimization, J. Opt. Theory Appl., 14, 31-49.

Bourbaki, N. [1968], Intégration, Livre V, Hermann, Paris.

Brauer, F. [1984], A class of differential-difference equations arising in delayed-recruitment population models, University of Wisconsin, Madison, Wis.

Brauer, F., Soudack, A.C. [1984], Optimal harvesting in predator-prey systems, Dept. of Mathematics, University of Wisconsin, Madison, Wis.

Brokate, M. [1980], A regularity condition for optimization in Banach spaces: Counter examples, Appl. Math. Optim., 6, 189-192.

Brunovsky, P. [1969], Controllability and linear closed-loop controls in linear periodic systems, J. Diff. Equns., 6, 296-313.

Buehler, H.H. [1975/6], Application of Neustadt's Theory of Extremals to an optimal Control Problem with a Functional Differential Equation and a Functional Inequality Constraint, Applied Math. Optim., 2, 34-74.

Butzer, P.L., Nessel, R.J. [1971], Fourier Analysis and Approximation, Vol. 1, Birkhäuser.

Cesari, L. [1983], Optimization - Theory and Applications, Problems with Ordinary Differential Equations, Springer Verlag.

Chan, W.L., Ng, S.K. [1979], Normality and proper performance improvement in periodic control, J. Opt. Theory Appl., 29, 215-229.

Chuang, C.-H., Speyer, J.L. [1985], Periodic optimal hypersonic scramjet cruise, Dept. Aerospace Engineering and Engineering Mechanics, University of Texas, Austin, Tx.

Clarke, F.H. [1976], The maximum principle under minimal hypotheses, SIAM J. Control Optim., 14, 1078-1091.

Clarke, F.H. [1983], Optimization and Nonsmooth Analysis, Wiley-Interscience.

Colonius, F. [1981], A penalty function proof of a Lagrange multiplier theorem with application to linear delay systems, Appl. Math. Optim. 7, 309-334.

Colonius, F. [1982a], Stable and regular reachability of relaxed hereditary differential systems, SIAM J. Control Optim., 20, 675-694. Addendum: ibid., 23 [1985].

Colonius, F. [1982b], The maximum principle for relaxed hereditary differential systems, SIAM J. Control Optim., 20, 695-712.

Colonius, F. [1982c], Optimal control of linear retarded system to small solutions, Int. J. Control, 36, 675-69 .

Colonius, F. [1983], A note on the existence of Lagrange multipliers, Appl. Math. Optim., 10, 187-191.

Colonius, F. [1984], An algebraic characterization of closed small attainability subspaces of delay systems, J. Diff. Equations, 53, 415-432.

Colonius, F. [1985a], The high frequency Pi-Criterion for retarded systems, IEEE Trans. Aut. Control, AC-30.

Colonius, F. [1985b], Optimal periodic control of retarded Liênard equations, Distributed Parameter Systems, F. Kappel, K. Kunisch, W. Schappacher, eds., Springer-Verlag, 77-91.

Colonius, F. [1986a], A note on optimal periodic control and nested optimization problems, J. Optim. Theory Appl., 50, 525-533.

Colonius, F. [1986b], Optimality for periodic control of functional differential systems, J. Math. Anal. Appl., 120, 119-149.

Colonius, F. [1987], Optimal periodic control of quasilinear systems in Hilbert spaces, Optimal Control of Partial Differential Equations II, K.H. Hoffmann, W. Krabs, eds., Birkhäuser, 57-66.

Colonius, F. [1988], Optimal periodic control: A scenario for local properness, SIAM J. Control Optim., 26 (1988).

Colonius, F., Hinrichsen, D. [1978], Optimal control of functional differential systems, SIAM J. Control Opt., 16, 861-879.

Colonius, F., Kliemann, W. [1986], Infinite time optimal control and periodicity, Institute for Mathematics and its Applications, University of Minnesota, Minneapolis.

Colonius, F., Manitius, A., Salamon, D. [1985], Structure theory and duality for time varying functional differential equations, to appear in J. Diff. Equations.

Colonius, F., Sieveking, M. [1987], Asymptotic properties of optimal solutions in planar discounted control problems, submitted to: SIAM J. Control Optim.

DaPrato, G. [1987], Synthesis of optimal control for an infinite dimensional periodic problem, SIAM J. Control Optim., 25, 706-714.

Das, P.C. [1975], Application of Dubovitskii/Milyutin formalism to optimal settling problems with constraints, Optimization and Optimal Control (Oberwolfach 1974), Springer Verlag.

Deklerk, M., Gatto, M. [1981], Some remarks on periodic harvesting of a fish population, Math. Biosciences, 56, 47-69.

Delfour, M.C. [1977], State theory of linear hereditary differential systems, J. Math. Anal. Appl., 60, 8-35.

Delfour, M.C., Karrakchou, J. [1987], State space theory of linear time invariant systems with delays in state, control and observation variables, I, J. Math. Anal. Appl., 125, 361-399, II, ibid., 400-450.

Delfour, M.C., Manitius, A. [1980], The structural operator F and its role in the theory of retarded systems, Part 1: J. Math. Anal. Appl. 73, 466-490, Part 2: J. Math. Anal. Appl. 74, 359-381.

De Pascale, E., Iannacci, R. [1983], Periodic solution of generalized Liênard equations with delay, Equadiff 82, H.W. Knobloch and K. Schmitt, eds., Springer-Verlag, 148-156.

Dickmanns, E.D. [1982], Optimaler periodischer Delphin-Segelflug, Forschungsschwerpunkt Simulation und Optimierung deterministischer und stochastischer Systeme, Hochschule der Bundeswehr, München.

Diekmann, O. [1980], Volterra integral equations and semigroups of operators, Preprint, Mathematisch Centrum, Report TW 197/80, Amsterdam.

Diekmann, O. [1981], A duality principle for delay equations, Preprint, Mathematisch Centrum Report TN 100/81, Amsterdam.

Diekmann, O., Van Gils, S.A. [1984], Invariant manifolds for Volterra integral equations of convolution type, J. Diff. Equations, 54, 139-180.

Diestel, J., Uhl, J.J. [1977], Vector Measures, AMS, Providence.

Dorato, P., Li, Y. [1985], A computer algorithm for a class of periodic optimization problems, presented at the American Control Conference, Boston.

Dorato, P., Knudsen, H.K. [1979], Periodic Optimization with Applications to Solar Energy Control, Automatica, 15, 673-676.

Douglas, J.M. [1972], Process Dynamics and Control, 1 & 2, Prentice Hall, Englewood Cliffs, New Jersey.

Dunford, N., Schwartz, J.T. [1967], Linear Operators, Part 1: General Theory, Wiley.

Edwards, R.E. [1967], Fourier Series, Vol. 1, Holt, Reinhart and Winston, New York.

Ekeland, I. [1974], On the variational principle, J. Math. Anal. Appl. 47, 324-353.

Ekeland, I. [1979], Nonconvex minimization problems, Bull. Amer. Math. Soc. (N.S.), 1, 443-474.

Fattorini, H.O. [1985], The maximum principle for nonlinear nonconvex systems in infinite dimensional spaces, Distributed Parameter Systems, F. Kappel, K. Kunisch, W. Schappacher, eds., Springer-Verlag, 162-178.

Fattorini, H.O. [1987], A unified theory of necessary conditions for nonlinear nonconvex control systems, Appl. Math. Optim., 15, 141-185.

Fiacco, A.V. [1976], Sensitivity analysis for nonlinear programming using penalty methods. Math. Programming 10, 287-311.

Fiacco, A.V. [1983], Introduction to Sensitivity and Stability Analysis in Nonlinear Programming, Academic Press.

Frêchet, M. [1915], Sur les fonctionelles bilinéaires, Trans. AMS 16, 215-234.

Gabasov, R., Kirillova, F. [1976], The Qualitative Theory of Optimal Processes, New York.

Gabasov, R., Kirillova, F.M. [1981], Mathematical theory of optimal control, J. Soviet Math., 17, 706-732.

Gaines, R.E., Peterson, J.K. [1983], Degree theoretic methods in optimal control, J. Math. Anal. Appl., 94, 44-77.

Gambaudo, J.M. [1985], Perturbation of a Hopf bifurcation by an external time-periodic forcing, J. Diff. Equations, 57, 172-199.

Gilbert, E.G. [1976], Vehicle Cruise: Improved fuel economy by periodic control, Automation 12, 159-166.

Gilbert, E.G. [1977], Optimal periodic control: A general theory of necessary conditions, SIAM J. Control Opt., 15, 717-746.

Gilbert, E.G. [1978], Optimal periodic control: A solution set theory of necessary and sufficient conditions, Preprints of the 7th IFAC Congress, Helsinki, pp. 2057-2064.

Gilbert, E.G., Lyons, D.T. [1981], The improvement of aircraft specific range by periodic control, AIAA Guid. and Control Conf., Albuquerque, N.M.

Gollan, B. [1981], Perturbation theory for abstract optimization problems, J. Opt. Theory Appl., 35, 417-442.

Golubitsky, M., Keyfitz, B.L. [1980], A qualitative study of the steady state solutions for a continuous flow stirred tank chemical reactor, SIAM J. Math. Anal., 11, 316-339.

Golubitsky, M., Schaeffer, D.G. [1985], Singularities and Groups in Bifurcation Theory, Springer-Verlag.

Guarbadassi, G. [1971], Optimal steady state versus periodic control, Ric. di Automatica, 2, 240-252.

Guarbadassi, G. [1976], The optimal periodic control problem, Journal A, 17, 75-83.

Guarbadassi, G., Locatelli, A., Rinaldi, S. [1974], Status of periodic optimization of Dynamical Systems, J. Opt. Theory Appl. 14, 1.20.

Guarbadassi, G., Schiavoni, N. [1975], Boundary optimal constant control versus periodic operation, 6th Triennial World Congress, IFAC Boston.

Halanay, A. [1974], Optimal control of periodic motions, Rev. Roum. Math. Pures Appl., 3-16.

Hale, J. [1977], Theory of Functional Differential Equations, Second Edition, Springer-Verlag.

Han, Mangasarian [1979], Exact penalty functions in mathematical programming, Math. Programming, 17, 251-269.

Hassard, B.D., Kazarinoff, N.D., Wang, Y.H. [1981], Theory and Application of Hopf Bifurcation, London Mathematical Society Lecture Notes Series 41, Cambridge University Press, Cambridge.

Henry, D. [1971], The adjoint of a linear functional differential equation and boundary value problems, J. Diff. Equations, 9, 55-66.

Hewer, G.A. [1975], Periodicity, detectability and the matrix Riccati equation, SIAM J. Control, 13, 1235-1251.

Hoffmann, K.H., Kornstaedt, H.J. [1978], Higher order necessary conditions in abstract mathematical programming, J. Opt. Theory Appl. 26, 533-568.

Horn, F.J.M., Bailey, J.E. [1968], An application of the theorem of relaxed control to the problem of increasing catalyst selectivity, J. Opt. Theory Appl. 2, 441-449.

Horn, F.J.M., Lin, R.C. [1967], Periodic processes: a variational approach, Ind. Eng. Chem., Process Des. Dev., 6, 1, 21-30.

Houlihan, S.C., Cliff, E.M., Kelley, H.J. [1982], Study of Chattering Cruise, J. Aircraft, 19, 119-124.

Hutson, V.M. [1977], A note on a boundary value problem for linear differential-difference equations of mixed type, J. Math. Anal. Appl. 61, 416-425.

Ioffe, A.D., Tikhomirov, V.M [1979], Theory of Extremal Problems, North-Holland Publ., (Russian Original 1974).

Jacobs, M.Q. [1972], An optimization problem for an nth order scalar neutral functional differential equation with functional side conditions, Delay and Functional Differential Equations and their Applications, K. Schmitt (ed.), Academic Press, 345-352.

Jacobs, M.Q., Kao, T.J. [1972], An optimum settling problem for time-lag systems, J. Math. Anal. Appl., 40, 687-707.

Jacobs, M.Q., Langenhop, C.E. [1976], Criteria for function space controllability of linear neutral systems, SIAM J. Control Optim., 14, 1009-1048.

Jacobs, M.Q., Pickel, W.C. [1978], Numerical methods for the solution of time optimal control problems for hereditary systems, Optimal Control and Differential Equations, A.B. Schwarzkopf, W.G. Kelley, S.B. Eliason, eds., Academic Press, 151-163.

Jakubczyk, B. [1978], A classification of attainable sets of linear differential- difference systems, Preprint 134, Institute of Mathematics, Polish Academy of Science, Warsaw.

Johnson, G.W. [1984], An unsymmetric Fubini Theorem, Amer. Math. Monthly 91, 131-133.

Jongen, H.Th., Jonker, P., Twilt, F. [1983], Nonlinear Optimization in R^n, I: Morse Theory, Chebyshev Approximation, Peter Lang Verlag, Frankfurt.

Kano, H., Nishimura, T. [1979], Periodic solution of matrix Riccati equations with detectability and stabilizability, Int. J. Control 29, 471-487.

Kamenskii, G.A., Myshkis, A.D. [1972], Boundary value problems with infinite defect, Diff. Equations, 7, 1612-1618.

Kappel, F. [1984], Linear autonomous functional differential equations in the state space C, Institut für Mathematik, Universität Graz, Graz, Report Nr. 34.

Khandelwal, D.N., Sharma, J., Ray, L.M. [1979], Optimal periodic maintenance of a machine, IEEE Trans. Aut. Control, AC-24, 513.

Kim, B.K., Bien, Z. [1981], On function target control of dynamic systems with delays in state and control under bounded state constraints, Int. J. Control, 33, 891-902.

Kirsch, A., Warth, W., Werner, J. [1978], Notwendige Optimalitätsbedingungen und ihre Anwendung, Springer-Verlag.

Klee, V. [1969], Separation and support properties of convex sets - a survey, Control Theory and the Calculus of Variations, A.V. Balakrishnan, ed., Academic Press, New York, 235-303.

Koivo, H.N., Koivo, A.J. [1978], Control and Estimation of systems with time delays, Distributed Parameter systems, W.H. Ray and D.G. Lainiotis, eds., Marcel Dekker, New York and Basel, 249-320.

Kojima, M. [1980], Strongly stable stationary solutions in nonlinear programs, Analysis and Computation of Fixed Points, S.M. Robinson, ed., Academic Press, New York.

Kowalsky, H.J [1963], Lineare Algebra, De Gruyter, Berlin.

Kurcyusz, S. [1973], A local maximum principle for operator constraints and its application to systems with time lag, Control and Cybernetics, 2, 99-125.

Kurcyusz, S. [1976], On the existence and nonexistence of Lagrange multipliers in Banach spaces, J. Optim. Theory Appl., 20, 81-110.

Kurcyusz, S, Olbrot, A.W. [1977], On the closure in $W^{1,q}$ of the attainable subspace of linear time lag systems, J. Diff. Equations, 24, 29-50.

Lasiecka, I., Manitius, A. [1985], Differentiability and convergence rates of approximating semigroups for retarded functional differential equations, Report, Rensselear Polytechnic Institute, Troy, N.Y.

Leizarowitz, A. [1985], Infinite horizon systems with unbounded cost, Appl. Math. Optim., 13, 19-43.

Lempio, F., Maurer, H. [1980], Differential stability in infinite-dimensional nonlinear programming, Appl. Math. Optim., 6, 139-152.

Lempio, F., Zowe, J. [1982], Higher order optimality conditions, Modern Applied Mathematics - Optimization and Operations Research, B. Korte, ed., North-Holland, 147-193.

Li, X.J. [1985], Maximum principle of optimal periodic control for functional differential systems, J. Optim. Theory Appl., 50, 421-429.

Li, X.J., Chow, S.N. [1987], Maximum principle of optimal control of functional differential systems, J. Optim. Theory Appl., 54, 335-360.

Linnemann, A. [1982], Higher-order necessary conditions for infinite and semi-infinite optimization, J.Optim.Theory Appl., 38, 483-512.

Lyusternik, L.A. [1934], Conditional Extrema of functionals, Mat. Sb., 41, 390-401.

Lorentz, R. [1978], Normalität der notwendigen Optimalitätsbedingungen bei einer Steueraufgabe mit Hammersteinscher Beschränkung, 6. Kolloquium "Optimierung dynamischer Systeme", Heringsdorf.

Luenberger, D.G. [1968], Optimization by Vector space methods, John Wiley, New York.

Maffezzoni, C. [1974], Hamilton-Jacobi theory for periodic control problems, J. Opt. Theory Appl. 14, 21-29.

Makowski, K., Neustadt, L.W. [1974], Optimal control problems with mixed control phase variable equality and inequality constraints, SIAM J. Control, 12, 184-228.

Malek-Zavarei, M., Jamshidi, M. [1987], Time Delay Systems, Analysis, Optimization and Applications, North-Holland 1987.

Manitius, A. [1974], Mathematical models of hereditary systems, Centre de Rêcherches Mathêmatiques, CRM-462, Universitê de Montrêal, Montreal.

Manitius, A. [1974], On the optimal control of systems with a delay depending on state, control and time, CRM-449, Université de Montrêal, also: Seminaires IRIA, Analyse et Contrôle de Systêmes, IRIA, France, 149-198.

Manitius, A. [1976], Optimal control of hereditary systems, Control Theory and Topics in Functional Analysis, International Atomic Energy Agency, Vienna.

Manitius, A. [1980], Completeness and F-Completeness of Eigenfunctions associated with retarded functional differential equations, J. Diff. Equ. 35., 1-29.

Manitius, A. [1981], Necessary and sufficient conditions of approximate controllability for general linear retarded systems, SIAM J. Control Opt., 19, 516-532.

Manitius, A. [1982], F-controllability and observability of linear retarded systems, Appl. Math. Opt., 9, 73-95.

Markus, L. [1973], Optimal control of limit cycles or what control theory can do to cure a heart attack or to cause one, Symposium on Ordinary Differential Equations, Minneapolis, Minnesota (1972), W.A. Harris, Y. Sibuya, eds., Springer-Verlag, Berlin.

Mathis, F.H., Reddien, G.W. [1978], Difference approximations to control problems with functional arguments, SIAM J. Control Optim., 16, 436-449.

Marshall, I.E. [1980], Control of Time Delay Systems, IEE Control Engineering Series 10, P. Peregrinus.

Marzollo, A. (ed.) [1972], Periodic Optimization, CISM Courses and Lectures No. 135, Springer-Verlag.

Matsubara, M., Nishimura, N., Watanabe, N., Onogi, K. [1981], Periodic Control Theory and Applications, Research Reports of Automatic Control Laboratory Vol. 28, Faculty of Engineering, Nagoya University.

Matsubara, M., Onogi, K. [1977], Structure analysis of periodically controlled nonlinear systems via the stroboscopic approach, IEEE Trans. Aut. Control, AC-22, 678-680.

Matsubara, M., Onogi, K. [1978a], Unstable suboptimal periodic control of a certain chemical reactor, IEEE Trans. Aut. Control, AC-23, 1111-1113.

Matsubara, M., Onogi, K. [1978b], Stabilized suboptimal periodic control of a chemical reactor. IEEE Trans. Aut. Control, AC-23, 1005-1008.

Maurer, H., Zowe, J. [1979], First and second order necessary and sufficient optimality conditions for infinite dimensional programming problems, Math. Programming, 16, 98-110.

Miller, R.K. [1974], Linear Volterra integro-differential equations as semigroups, Funkcial, EKvac. 17, 39-55.

Miller, R.K., Michel, A.N. [1980], On existence of periodic motions in nonlinear control systems with periodic inputs, SIAM J. Control Opt. 18, 585-598.

Morse, M. [1950], Bilinear functionals over C×C, Acta Scientarum Mathematicarum, XII, Pars B, 41-48, Szeged.

Neustadt, L.W. [1963], The existence of optimal controls in the absence of convexity conditions, J. Math. Anal. Appl. 7, 110-117.

Neustadt, L.W. [1976], Optimization - A Theory of Necessary Conditions, Princeton University Press, Princeton.

Nistri, P. [1983], Periodic control problems for a class of nonlinear periodic differential systems, Nonlinear Analysis, Theory, Methods and Applications, 7, 79-90.

Noldus, E. [1975], A survey of optimal periodic control of continuous systems, Journal A, 16, 11-16.

Oguztöreli, M.N. [1966], Time-Lag Control Systems, Academic Press, New York.

Olbrot, A.W. [1976], Control of retarded systems with function space constraints: necessary optimality conditions, Control Cybernet., 5, 5-31.

Olbrot, A.W. [1977], Some results on infinite-dimensional extremal problems, Theory of Nonlinear Operators, Constructive Aspects, R. Kluge, W. Müller, eds., Akademie Verlag, Berlin (GDR).

Onogi, K., Matsubara, M. [1980], Structure Analysis of periodically controlled chemical processes, Chem. Eng. Sci., 34, 1009-1019.

Ortlieb, C.P. [1980], Optimale periodische Steuerung diskreter Prozesse, Constructive methods of finite nonlinear Optimization, L. Collatz, G. Meinardus, W. Wetterling, eds., Birkhäuser, Basel, 179-196.

Pazy, A. [1983], Semigroups of linear Operators and Applications to Partial Differential Equations, Springer-Verlag.

Penot, J.P. [1982], On regularity conditions in mathematical programming, Mathematical Programming Study 19, 167-199.

Pontryagin, L.S., Boltyanskii, V.G., Gamkrelidze, R.V., Mischenko, E.F. [1962], Mathematical Theory of Optimal Processes, Interscience Publ., New York, N.Y.

Poore, A.B. [1974], A model equation arising from chemical reactor theory, Arch. Rational Mech. Anal., 52, 358-388.

Rabinowitz, P.H. [1982], Periodic solutions of Hamiltonian Systems: A survey, SIAM J. Math. Anal., 13, 343-342.

Ray, W.H., Soliman, M.A. [1970], The optimal control of processes containing pure time delays - I: Necessary conditions for an optimum, Chem. Eng. Sci., 25, p. 1911.

Reddien, G.W., Travis, C.C. [1974], Approximation methods for boundary value problems of differential equations with functional arguments, J. Math. Anal. Appl. 46, 62-74.

Robinson, S.M. [1976], Stability theory for systems of inequalities in nonlinear programming, part II: differentiable nonlinear systems, SIAM J. Num. Anal. 13, 497-513.

Rockey, S.A. [1982], Discrete Methods in State Approximation, Parameter Identification and Optimal Control for Hereditary Systems, Ph.D. Thesis, Brown University, Providence, R.I.

Rodas, H.R., Langenhop, C.E. [1978], A sufficient condition for function space controllability of a linear neutral system, SIAM J. Control Optim. 16, 429-435.

Roxin, E.D., Stern, L.E. [1982], Periodicity in optimal control and differential games, System Modelling and Optimization, R.F. Drenick and F. Kozin, eds., Springer-Verlag.

Russell, D.L. [1982], Optimal orbital regulation in dynamical systems subject to Hopf bifurcation, J. Diff. Equ. 44, 188-223.

Sachs, G., Christodopulou [1986], Reducing fuel consumption by cyclic control, ICAS 86-5.1.2, Technische Universität München, München.

Salamon, D. [1982], Control and Observation of Neutral Systems, Dissertation, Universität Bremen, Bremen, = Research Notes in Mathematics, Vol. 91, Pitman, Boston-London-Melbourne 1984.

Salamon, D. [1985], Structure and stability of finite dimensional approximations for functional differential equations, SIAM J. Control Optim., 23, 928-951.

Schädlich, K., Hoffmann, U., Hofmann, H. [1983], Periodical operation of chemical processes and evaluation of conversion improvements, Chemical Engineering Science, 38, 1375-1384.

Seinfeld, J.H. [1969], Optimal control of a continuous stirred tank reactor with transportation lag, Int. J. Control, 10, 29-39.

Sincic, D., Bailey, J.E. [1977], Pathological dynamic behaviour of forced periodic chemical processes, Chem. Eng. Sci., 32, 281-286.

Sincic, D., Bailey, J.E. [1978], Optimal periodic control of variable time-delay systems, Int. J. Control, 27, 547-555.

Sincic, D., Bailey, J.E. [1980], Analytical Optimization and Sensitivity analysis of forced periodic chemical processes, Chem. Eng. Sci. 35, 1153-1165.

Soliman, M.A., Ray, W.H. [1972a], On the optimal control of systems having pure time delays and singular arcs, I, Necessary conditions for optimality, Int. J. Control, 16, No. 5.

Soliman, M.A., Ray, W.H. [1972b], Optimal control of multivariable systems with pure time delays, Automatica, 7, 681-689.

Speyer, J.L. [1973], On the fuel optimality of cruise, Journal of Aircraft, 10, 763-764.

Speyer, J.L. [1976], Non-optimality of steady-state cruise for aircraft, AIAA Journal, 14, 1604-1610.

Speyer, J.L., Dannemiller, D., Walker, D. [1985], Periodic optimal cruise of an athmospheric vehicle, J. Guidance, Control and Dynamics, 8, 31-38.

Speyer, J.L., Evans, R.T. [1984], A second variational theory of optimal periodic processes, IEEE Trans. Aut. Control, 29, 138-148.

Timonen, J., Hämäläinen, R.P. [1979], Optimal periodic control strategies in a dynamic pricing problem, Int. J. Systems Sci., 10, 197-205.

Tzafestas, S.G. [1983], Walsh transform theory and its application to systems analysis and control: an overview, Mathematics and Computers in Simulation XXV, North-Holland, 214-225.

Uppal, A., Ray, W.H., Poore, A.B. [1974], On the dynamic behaviour of continuous stirred tank reactors, Chem. Eng. Sci., 29, 967-985.

Utthoff, J. [1979], Optimale Kontrolle neutraler Funktional-Differentialgleichungen, Diplomarbeit, Freie Universität Berlin, Berlin.

Vainberg, M.M. [1964], Variational Methods for the Study of Nonlinear Operators, Holden-Day, San Francisco.

Vainberg, M.M. [1952], Some problems in the differential calculus in linear spaces, Uspehi Mat. Nauk, 7, No.4, 55-102 (in Russian).

Valkó, P., Almasy, G.A. [1982], Periodic optimization of Hammerstein-type systems, Automatica, 18, 245-248.

Vincent, T.L., Lee, C.S., Goh, B.S. [1977], Control targets for the management of biological systems, Ecol. Modelling, 3, 285-300.

Vogel, J. [1979], Untersuchungen zur Normalität bei Aufgaben der optimalen Steuerung, 24, Intern. Wiss. Kolloq., TH Ilmenau.

Warga, J. [1971], Normal control problems have no minimizing strictly original solutions, Bull. Amer. Math. Soc., 77, 625-628.

Warga, J. [1972], Optimal Control of Differential and Functional Equations, Academic Press, New York.

Warga, J. [1974], Optimal controls with pseudo-delays, SIAM J. Control, 12, 286-299.

Watanabe, N., Nishimura, Y., Matsubara, M. [1976], Singular control test for optimal periodic control problems, IEEE Trans. Aut. Control, AC-21, 609-610.

Watanabe, N., Onogi, K., Matsubara, M. [1981], Periodic control of continuous stirred tank reactors - I, Chem. Eng. Sci., 36, 809-818, II idid. 37, 745-752.

Watanabe, N., Kurimoto, H., Matsubara, M. [1984], Periodic control of continuous stirred tank reactors - III, Case of multistage reactors, Chem. Eng. Sci., 39, 31-36.

Werner, J. [1984], Optimization Theory and Applications, Vieweg, Braunschweig/Wiesbaden.

Wierzbicki, A.P., Hatko, A. [1973], Computational methods in Hilbert space for optimal control problems with delays, 5th IFIP Conf. Optimization Techniques, R. Conti, A. Ruberti, eds., Springer-Verlag.

Williamson, L.J., Polak, E. [1976], Relaxed controls and the convergence of optimal control algorithms, SIAM J. Control Optim., 14, 737-756.

Young, L.C. [1980], Lectures on the Calculus of Variations and Optimal Control Theory (2nd ed.), Chelsea Publ. Co., New York.

Zowe, J., Krucyusz, S. [1979], Regularity and Stability for the mathematical programming problem in Banach spaces, Appl. Math. Optim., 5, 49-62.

MIX
Papier aus verantwortungsvollen Quellen
Paper from responsible sources
FSC® C105338

If you have any concerns about our products,
you can contact us on
ProductSafety@springernature.com

In case Publisher is established outside the EU,
the EU authorized representative is:
Springer Nature Customer Service Center GmbH
Europaplatz 3, 69115 Heidelberg, Germany

Printed by Libri Plureos GmbH
in Hamburg, Germany